Friedrich Wilhelm Georg Kohlrausch, And others

The Fundamental Laws of Electrolytic Conduction

Memoirs by Faraday, Hittorf and F. Kohlrausch

Friedrich Wilhelm Georg Kohlrausch, And others

The Fundamental Laws of Electrolytic Conduction
Memoirs by Faraday, Hittorf and F. Kohlrausch

ISBN/EAN: 9783337232207

Printed in Europe, USA, Canada, Australia, Japan

Cover: Foto ©berggeist007 / pixelio.de

More available books at **www.hansebooks.com**

THE FUNDAMENTAL LAWS

OF

ELECTROLYTIC CONDUCTION

MEMOIRS BY FARADAY, HITTORF
AND F. KOHLRAUSCH

TRANSLATED AND EDITED

By H. M. GOODWIN, Ph.D.

ASSISTANT PROFESSOR OF PHYSICS
MASSACHUSETTS INSTITUTE OF TECHNOLOGY

NEW YORK AND LONDON
HARPER & BROTHERS PUBLISHERS
1899

GENERAL CONTENTS

PREFACE

In the present volume are collected those papers on electrochemistry which contain the original statement of the fundamental laws and experiments on which the modern theory of electrolytic conduction is based. Of these, Faraday's law of definite electrochemical action and electrochemical equivalents, first stated in 1834, naturally takes precedence. This law is universally recognized as one of the few rigidly exact laws of nature, and lies at the basis of all electrochemical theory and practice. Of the extended series of experiments in electrochemistry, contained in the fifth and seventh series of Faraday's *Experimental Researches*, all of which touch more or less on the law in question, only those sections which have a direct bearing on the establishment of the law are here presented. Faraday's brief paper on the "Relation by Measure of Common and Voltaic Electricity" has been added as an introduction, as it was in this article that he was first led to a statement of the probable existence of the law to which he afterwards devoted so much attention.

Second only to Faraday's law, the classical researches of Hittorf on the concentration changes produced at the electrodes during electrolysis, have proved of fundamental significance in the explanation of electrolytic phenomena. The explanation given by Hittorf in 1853 of this phenomenon is still that generally accepted by physicists at the present time. Of Hittorf's five papers bearing on this subject, all of which are easily accessible in German in Ostwald's *Klassiker der Exakten Wissenschaften*, the first only has been here translated. This, however, is complete in itself, and contains not only a statement of Hittorf's theory, but also a comprehensive and remarkably careful experimental investigation of the phenomenon of transference. The later papers are mainly an exten-

PREFACE

sion of the first, with applications to certain important problems of chemical constitution.

The great importance of the results obtained by Hittorf was not generally recognized at the time of their publication, but only after F. Kolrausch had pointed out their bearing on his investigations on the electrical conductivity of solutions. The elegance of method and accuracy with which these investigations have been, and are still being carried out, place them pre-eminent among investigations of this class. Immediately after sufficient conductivity data had been obtained, Kohlrausch recognized the bearing of Hittorf's investigations upon his results, and was led to the formulation of the law of the independent migration of ions. The paper in which this law was first presented to the Göttingen Academy in 1876 is translated in full. It was not until 1879 that the researches, of which this was the most important conclusion, appeared in complete form in Wiedemann's *Annalen*.

With the establishment of the laws of Faraday, Hittorf, and Kohlrausch the way was prepared for the dissociation theory of Arrhenius, which was announced in 1886, as soon as the theory of solutions had been formulated by Van't Hoff.

H. M. Goodwin.

Massachusetts Institute of Technology.

RELATION BY MEASURE

OF

COMMON AND VOLTAIC ELECTRICITY

BY

MICHAEL FARADAY

Read January 17, 1833, before the Royal Society

(*Philosophical Transactions*, **123**, 48, 1833 ; Poggendorff's *Annalen*, **29**, 373, 1833 ; *Experimental Researches in Electricity*, Vol. I., Series III., § 8, p. 102)

CONTENTS

RELATION BY MEASURE

OF

COMMON AND VOLTAIC ELECTRICITY

BY

MICHAEL FARADAY

BELIEVING the point of identity to be satisfactorily established,* I next endeavored to obtain a common measure, or a known relation as to quantity, of the electricity excited by a machine, and that from a voltaic pile; for the purpose not only of confirming their identity, but also of demonstrating certain general principles and creating an extension of the means of investigating and applying the chemical powers of this wonderful and subtile agent.

The first point to be determined was, whether the same absolute quantity of ordinary electricity, sent through a galvanometer, under different circumstances, would cause the same deflection of the needle. An arbitrary scale was therefore attached to the galvanometer, each division of which was equal to about 4°, and the instrument arranged as in former experiments. The machine, battery, and other parts of the apparatus were brought into good order, and retained for the time as nearly as possible in the same condition. The experiments were alternated so as to indicate any change in the condition of the apparatus and supply the necessary corrections.

Seven of the battery jars were removed and eight retained for present use. It was found that about forty turns would

---

* [*In the paper immediately preceding this, the "Identity of Electricities from Different Sources" was experimentally demonstrated for voltaic, frictional, magneto, thermo, and animal electricity.*]

3

fully charge the eight jars. They were then charged by thirty turns of the machine, and discharged through the galvanometer, a thick wet string, about ten inches long, being included in the circuit. The needle was immediately deflected five divisions and a half, on the one side of the zero, and in vibrating passed as nearly as possible through five divisions and a half on the other side.

The other seven jars were then added to the eight, and the whole fifteen charged by thirty turns of the machine. The Henley electrometer stood not quite half as high as before; but when the discharge was made through the galvanometer, previously at rest, the needle immediately vibrated, passing *exactly* to the same division as in the former instance. These experiments with eight and fifteen jars were repeated several times alternately with the same results.

Other experiments were then made, in which all the battery was used, and its charge (being fifty turns of the machine) sent through the galvanometer : but it was modified by being passed sometimes through a mere wet thread, sometimes through thirty-eight inches of thin string wetted by distilled water, and sometimes through a string of twelve times the thickness, only twelve inches in length, and soaked in dilute acid. With the thick string the charge passed at once ; with the thin string it occupied a sensible time ; and with the thread it required two or three seconds before the electrometer fell entirely down. The current therefore must have varied extremely in intensity in these different cases, and yet the deflection of the needle was sensibly the same in all of them. If any difference occurred, it was that the thin string and thread caused greatest deflection ; and if there is any lateral transmission, as M. Colladon says, through the silk in the galvanometer coil, it ought to have been so, because then the intensity is lower and the lateral transmission less.

Hence it would appear that *if the same absolute quantity of electricity passes through the galvanometer, whatever may be its intensity, the deflecting force upon the magnetic needle is the same.*

The battery of fifteen jars was then charged by sixty revolutions of the machine, and discharged, as before, through the galvanometer. The deflection of the needle was now as nearly as possible to the eleventh division, but the graduation was not accurate enough for me to assert that the arc was exactly double the former arc ; to the eye it appeared to be so. The probability is that *the deflecting force of an electric current is directly proportional to the absolute quantity of electricity passed,* at whatever intensity that electricity may be.*

Dr. Ritchie has shown that in a case where the intensity of the electricity remained the same, the deflection of the magnetic needle was directly as the quantity of electricity passed through the galvanometer.† Mr. Harris has shown that the *heating* power of common electricity on metallic wires is the same for the same quantity of electricity, whatever its intensity might have previously been.‡

The next point was to obtain a *voltaic* arrangement producing an effect equal to that just described. A platina and a zinc wire were passed through the same hole of a draw-plate, being then one-eighteenth of an inch in diameter; these were fastened to a support, so that their lower ends projected, were parallel, and five-sixteenths of an inch apart. The upper ends were well connected with the galvanometer wires. Some acid was diluted, and, after various preliminary experiments, that was adopted as a standard which consisted of one drop strong sulphuric acid in four ounces of distilled water. Finally, the time was noted which the needle required in swinging either from right to left or left to right : it was equal to seventeen beats of my watch, the latter giving one hundred and fifty in a minute. The object of these preparations was to arrange a voltaic apparatus which, by immersion in a given acid for a given time, much less than that required by the needle to swing in one direction, should give equal deflection to the instrument with the discharge of ordinary electricity from the battery ; and a new part of the zinc wire having been brought into position with the platina, the comparative experiments were made.

---

* The great and general value of the galvanometer as an actual measure of the electricity passing through it, either continuously or interruptedly, must be evident from a consideration of these two conclusions. As constructed by Professor Ritchie, with glass threads (see *Philosophical Transactions,* 1830, p. 218, and *Quarterly Journal of Science,* New Series, vol. i., p. 29), it apparently seems to leave nothing unsupplied in its own department.

† *Quarterly Journal of Science,* New Series, vol. i., p. 33.

‡ *Plymouth Transactions,* p. 22.

On plunging the zinc and platina wires five-eighths of an inch deep into the acid, and retaining them there for eight beats of the watch (after which they were quickly withdrawn), the needle was deflected, and continued to advance in the same direction some time after the voltaic apparatus had been removed from the acid. It attained the five-and-a-half division. and then returned, swinging an equal distance on the other side. This experiment was repeated many times, and always with the same result.

Hence, as an approximation, and judging from *magnetic force* only at present, it would appear that two wires, one of platina and one of zinc, each one-eighteenth of an inch in diameter, placed five-sixteenths of an inch apart and immersed to the depth of five-eighths of an inch in acid, consisting of one drop oil of vitrol and four ounces distilled water, at a temperature of about 60°, and connected at the other extremities by a copper wire eighteen feet long and one-eighteenth of an inch thick (being the wire of the galvanometer coils), yield as much electricity in eight beats of my watch, or in $\frac{8}{180}$ of a minute, as the electrical battery charged by thirty turns of the large machine in excellent order. Notwithstanding this apparently enormous disproportion, the results are perfectly in harmony with those effects which are known to be produced by variations in the intensity and quantity of the electric fluid.

In order to procure a reference to *chemical action*, the wires were now retained immersed in the acid to the depth of five-eighths of an inch, and the needle, when stationary, observed ; it stood, as nearly as the unassisted eye could decide, at $5\frac{1}{4}$ division. Hence a permanent deflection to that extent might be considered as indicating a constant voltaic current, which in eight beats of my watch could supply as much electricity as the electrical battery charged by thirty turns of the machine.

The following arrangements and results are selected from many that were made and obtained relative to chemical action. A platina wire, one-twelfth of an inch in diameter, weighing two hundred and sixty grains, had the extremity rendered plain, so as to offer a definite surface equal to a circle of the same diameter as the wire; it was then connected in turn with the conductor of the machine or with the voltaic apparatus, so as always to form the positive pole, and at the same time retain a perpendicular position, that it might rest with its whole weight upon the test paper to be employed. The test paper itself was supported upon a platina spatula, connected either with a discharging train or with the negative wire of the voltaic apparatus, and it consisted of four thicknesses, moistened at all times to an equal degree in a standard solution of hydriodate of potassa.

When the platina wire was connected with the prime conductor of the machine and the spatula with the discharging train, ten turns of the machine had such decomposing power as to produce a pale, round spot of iodine of the diameter of the wire ; twenty turns made a much darker mark, and thirty turns made a dark-brown spot, penetrating to the second thickness of the paper. The difference in effect produced by two or three turns, more or less, could be distinguished with facility.

The wire and the spatula were then connected with the voltaic apparatus, the galvanometer being also included in the arrangement; and, a stronger acid having been prepared, consisting of nitric acid and water, the voltaic apparatus was immersed so far as to give a permanent deflection of the needle to the 5½ division, the fourfold moistened paper intervening as before.* Then by shifting the end of the wire from place to place upon the test paper, the effect of the current for five, six, seven, or any number of beats of the watch was observed and compared with that of the machine. After alternating and repeating the experiments of comparison many times, it was constantly found that this standard current of voltaic electricity, continued for eight beats of the watch, was equal in chemical effect to thirty turns of the machine ; twenty-eight revolutions of the machine were sensibly too few.

Hence it results that both in *magnetic deflection* and in *chemical force*, the current of electricity of the standard voltaic battery for eight beats of the watch was equal to that of the machine evolved by thirty revolutions.

It also follows that for this case of electro-chemical decomposition, and it is probable for all cases, that the *chemical power, like the magnetic force, is in direct proportion to the absolute quantity of electricity* which passes.

Hence arises still further confirmation, if any were required,

* Of course the heightened power of the voltaic battery was necessary to compensate for the bad conductor now interposed.

7

of the identity of common and voltaic electricity, and that the differences of intensity and quantity are quite sufficient to account for what were supposed to be their distinctive qualities.

The extension which the present investigations have enabled me to make of the facts and views constituting the theory of electrochemical decomposition will, with some other points of electrical doctrine, be almost immediately submitted to the Royal Society in another series of the Researches.

ROYAL INSTITUTION, *December* 15, 1832

ON ELECTROCHEMICAL DECOMPOSITION

BY

MICHAEL FARADAY

Read January 23, February 6 and 13, 1834, before the Royal Society

(*Philosophical Transactions*, **124**, 77, 1834; Poggendorff's *Annalen*, **33**, 301, 1834; *Experimental Researches in Electricity*, Vol. I., Series VII., § 11, p. 195)

CONTENTS

ON ELECTROCHEMICAL DECOMPOSITION

BY

MICHAEL FARADAY

PRELIMINARY

THE theory which I believe to be a true expression of the facts of electrochemical decomposition, and which I have therefore detailed in a former series of these Researches, is so much at variance with those previously advanced that I find the greatest difficulty in stating results, as I think, correctly, whilst limited to the use of terms which are current with a certain accepted meaning. Of this kind is the term *pole*, with its prefixes of positive and negative, and the attached ideas of attraction and repulsion. The general phraseology is that the positive pole *attracts* oxygen, acids, etc., or more cautiously, that it *determines* their evolution upon its surface ; and that the negative pole acts in an equal manner upon hydrogen, combustibles, metals, and bases. According to my view, the determining force is *not* at the poles, but *within* the body under decomposition ; and the oxygen and acids are rendered at the *negative* extremity of that body, whilst hydrogen, metals, etc., are evolved at the *positive* extremity.

To avoid, therefore, confusion and circumlocution, and for the sake of greater precision of expression than I can otherwise obtain, I have deliberately considered the subject with two friends, and with their assistance and concurrence in framing them, I purpose henceforward using certain other terms, which I will now define. The *poles*, as they are usually called, are only the doors or ways by which the electric current passes into and out of the decomposing body ; and they of course, when in contact with that body, are the limits of its extent in the direction of the current. The term has been generally applied

11

to the metal surfaces in contact with the decomposing substance; but whether philosophers generally would also apply it to the surfaces of air and water, against which I have effected electrochemical decomposition, is subject to doubt. In place of the term pole, I propose using that of *electrode*,* and I mean thereby that substance, or rather surface, whether of air, water, metal, or any other body, which bounds the extent of the decomposing matter in the direction of the electric current.

The surfaces at which, according to common phraseology, the electric current enters and leaves a decomposing body are most important places of action, and require to be distinguished apart from the poles, with which they are mostly, and the electrodes, with which they are always, in contact. Wishing for a natural standard of electric direction to which I might refer these, expressive of their difference and at the same time free from all theory, I have thought it might be found in the earth. If the magnetism of the earth be due to electric currents passing round it, the latter must be in constant direction, which, according to the present usage of speech, would be from east to west, or, which will strengthen this to help the memory, that in which the sun appears to move. If in any case of electro-decomposition we consider the decomposing body as placed so that the current passing through it shall be in the same direction, and parallel to that supposed to exist in the earth, then the surfaces at which the electricity is passing into and out of the substance would have an invariable reference, and exhibit constantly the same relations of powers. Upon this notion we purpose calling that towards the east the *anode*,† and that towards the west the *cathode*;‡ and whatever changes may take place in our views of the nature of electricity and electrical action, as they must affect the *natural standard* referred to, in the same direction, and to an equal amount with any decomposing substances to which these terms may at any time be applied, there seems no reason to expect that they will lead to confusion or tend in any way to support false views. The *anode* is therefore that surface at which the electric current, according to our present expression, enters: it is the *negative* extremity of the decomposing body; is where oxygen,

---

* ἤλεκτρον, and ὁδὸς, *a way.*

† ἄνω, *upwards*, and ὁδὸς, *a way*: the way which the sun rises.

‡ κατὰ, *downwards*, and ὁδὸς, *a way*: the way which the sun sets.

chlorine, acids, etc., are evolved; and is against or opposite the positive electrode. The *cathode* is that surface at which the current leaves the decomposing body, and is its *positive* extremity; the combustible bodies, metals, alkalies, and bases, are evolved there, and it is contact with the negative electrode. I shall have occasion in these Researches, also, to class bodies together according to certain relations derived from their electrical actions; and wishing to express those relations without at the same time involving the expression of any hypothetical views, I intend using the following names and terms. Many bodies are decomposed directly by the electric current, their elements being set free; these I propose to call *electrolytes.** Water, therefore, is an electrolyte. The bodies which, like nitric or sulphuric acids, are decomposed in a secondary manner are not included under this term. Then for *electrochemically decomposed*, I shall often use the term *electrolyzed*, derived in the same way, and implying that the body spoken of is separated into its components under the influence of electricity: it is analogous in its sense and sound to *analyze*, which is derived in a similar manner. The term *electrolytical* will be understood at once: muriatic acid is electrolytical, boracic acid is not.

Finally, I require a term to express those bodies which can pass to the *electrodes*, or, as they are usually called, the poles. Substances are frequently spoken of as being *electro-negative* or *electro-positive*, according as they go under the supposed influence of a direct attraction to the positive or negative pole. But these terms are much too significant for the use to which I should have to put them; for, though the meanings are perhaps right, they are only hypothetical, and may be wrong; and then, through a very imperceptible, but still very dangerous, because continual, influence, they do great injury to science by contracting and limiting the habitual views of those engaged in pursuing it. I propose to distinguish such bodies by calling those *anions*† which go to the *anode* of the decomposing body; and those passing to the *cathode, cations*‡; and when I have occasion to speak of these together, I shall call them *ions.* Thus, the chloride of lead is an *electrolyte*, and when *electrolyzed*

---

* ἤλεκτρον, and λύω, *solvo.* Noun, electrolyte; verb, electrolyze.

† ἀνιών, *that which goes up.* (Neuter participle.)

‡ κατιών, *that which goes down.*

13

evolves the two *ions*, chlorine and lead, the former being an *anion*, and the latter a *cation*.

These terms, being once well defined, will, I hope, in their use enable me to avoid much periphrasis and ambiguity of expression. I do not mean to press them into service more frequently than will be required, for I am fully aware that names are one thing and science another.*

It will be well understood that I am giving no opinion respecting the nature of the electric current now, beyond what I have done on former occasions ; and that though I speak of the current as proceeding from the parts which are positive to those which are negative, it is merely in accordance with the conventional, though in some degree tacit, agreement entered into by scientific men, that they may have a constant, certain, and definite means of referring to the direction of the forces of that current.

[*Section IV., including eight pages " On Some General Conditions of Electrochemical Decomposition," is here omitted.*]

ON A NEW MEASURER OF VOLTA-ELECTRICITY

I have already said, when engaged in reducing common and voltaic electricity to one standard of measurement,† and again when introducing my theory of electrochemical decomposition, that the chemical decomposing action of a current *is constant for a constant quantity of electricity*, notwithstanding the greatest variations in its sources, in its intensity, in the size of the *electrodes* used, in the nature of the conductors (or non-conductors) through which it is passed, or in other circumstances. The conclusive proofs of the truth of these statements shall be given almost immediately.

I endeavored upon this law to construct an instrument which should measure out the electricity passing through it, and which, being interposed in the course of the current used in any particular experiment, should serve at pleasure, either as a *comparative standard* of effect or as a *positive measurer* of this subtile agent.

---

* Since this paper was read, I have changed some of the terms which were first proposed, that I might employ only such as were at the same time simple in their nature, clear in their reference, and free from hypothesis.

† [*See page 7.*]

There is no substance better fitted, under ordinary circumstances, to be the indicating body in such an instrument than water; for it is decomposed with facility when rendered a better conductor by the addition of acids or salts ; its elements may in numerous cases be obtained and collected without any embarrassment from secondary action, and, being gaseous, they are in the best physical condition for separation and measurement. Water, therefore, acidulated by sulphuric acid, is the substance I shall generally refer to, although it may become expedient in peculiar cases or forms of experiment to use other bodies.

The first precaution needful in the construction of the instrument was to avoid the recombination of the evolved gases, an effect which the positive electrode has been found so capable of producing. For this purpose various forms of decomposing apparatus were used. The first consisted of straight tubes, each containing a plate and wire of platina soldered together by gold, and fixed hermetically in the glass at the closed extremity of the tube (Fig. 1). The tubes were about 8 inches long, 0.7 of an inch in diameter, and graduated. The platina plates were about an inch long, as wide as the tubes would permit, and adjusted as near to the mouths of the tubes as was consistent with the safe collection of the gases evolved. In certain cases, where it was required to evolve the elements upon as small a surface as possible, the metallic extremity, instead of being a plate, consisted of the wire bent into the form of a ring (Fig. 2). When these tubes were used as measurers, they were filled with dilute sulphuric acid, inverted in a basin of the same liquid

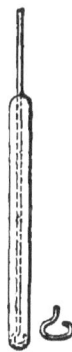

Fig. 1

Fig. 2

(Fig. 3), and placed in an inclined position, with their mouths near to each other, that as little decomposing matter should intervene as possible ; and also, in such a direction that the platina plates should be in vertical planes.

Another form of apparatus is that delineated (Fig. 4). The tube is bent in the middle ; one end is closed ; in that end is fixed

Fig. 3

a wire and plate, *a*, proceeding so far downwards, that, when

in the position figured, it shall be as near to the angle as possible, consistently with the collection at the closed extremity of the tube, of all the gas evolved against it. The plane of this plate is also perpendicular. The other metallic termination, *b*, is introduced at the time decomposition is to be effected, being brought as near the angle as possible, without causing any gas to pass from it towards the closed end of the instrument. The gas evolved against it is allowed to escape.

Fig. 4

The third form of apparatus contains both electrodes in the same tube; the transmission, therefore, of the electricity and the consequent decomposition is far more rapid than in the separate tubes. The resulting gas is the sum of the portions evolved at the two electrodes, and the instrument is better adapted than either of the former as a measurer of the quantity of voltaic electricity transmitted in ordinary cases. It consists of a straight tube (Fig. 5) closed at the upper extremity and graduated, through the sides of which pass platina wires (being fused into the glass), which are connected with two plates within. The tube is fitted by grinding into one mouth of a double-necked bottle. If the latter be one-half or two-thirds full of the dilute sulphuric acid, it will, upon inclination of the whole, flow into the tube and fill it. When an electric current is passed through the instrument, the gases evolved against the plates collect in the upper portion of the tube, and are not subject to the recombining power of the platina.

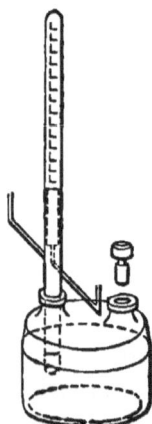

Fig. 5

Another form of the instrument is given in Fig. 6. A fifth form is delineated (Fig. 7). This I have found exceedingly useful in experiments continued in succession for days together,

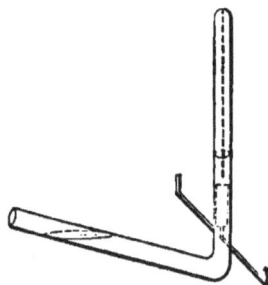

Fig. 6

and where large quantities of indicating gas were to be collected. It is fixed on a weighted foot, and has the form of a

16

small retort containing the two electrodes; the neck is narrow, and sufficiently long to deliver gas issuing from it into a jar placed in a small pneumatic trough. The electrode chamber, sealed hermetically at the part held in the stand, is 5 inches in length and 0.6 of an inch in diameter; the neck is about 9 inches in length and 0.4 of an inch in diameter internally. The figure will fully indicate the construction.

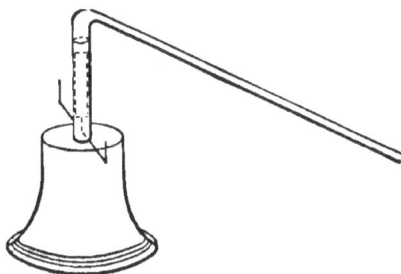

Fig. 7

It can hardly be requisite to remark, that in the arrangement of any of these forms of apparatus, they, and the wires connecting them with the substance, which is collaterally subjected to the action of the same electric current, should be so far insulated as to insure a certainty that all the electricity which passes through the one shall be transmitted through the other.

Next to the precaution of collecting the gases, if mingled, out of contact with the platinum, was the necessity of testing the law of a *definite electrolytic* action, upon water at least, under all varieties of condition; that, with a conviction of its certainty, might also be obtained a knowledge of those interfering circumstances which would require to be practically guarded against.

The first point investigated was the influence or indifference of extensive variations in the size of the electrodes, for which purpose instruments like those last described were used (Figs. 5, 6, and 7). One of these had plates 0.7 of an inch wide and nearly 4 inches long; another had plates only 0.5 of an inch wide and 0.8 of an inch long; a third had wires 0.02 of an inch in diameter and 3 inches long; and a fourth, similar wires only half an inch in length. Yet when these were filled with dilute sulphuric acid, and, being placed in succession, had one common current of electricity passed through them, very nearly the same quantity of gas was evolved in all. The difference was sometimes in favor of one, and sometimes on the side of another; but the general result was that the largest quantity of gases was evolved at the smallest electrodes — namely, those consisting merely of platina wires.

Experiments of a similar kind were made with the single-plate straight tubes (Figs. 1, 2, 3), and also with the curved tubes (Fig. 4), with similar consequences; and when these, with the former tubes, were arranged together in various ways, the result, as to the equality of action of large and small metallic surfaces when delivering and receiving the same current of electricity, was constantly the same. As an illustration, the following numbers are given. An instrument with two wires evolved 74.3 volumes of mixed gases; another with plates, 73.25 volumes; whilst the sum of the oxygen and hydrogen in two separate tubes amounted to 73.65 volumes. In another experiment the volumes were 55.3, 55.3, and 54.4.

But it was observed in these experiments, that in single-plate tubes more hydrogen was evolved at the negative electrode than was proportionate to the oxygen at the positive electrode; and generally, also, more than was proportionate to the oxygen and hydrogen in a double-plate tube. Upon more minutely examining these effects, I was led to refer them, and also the differences between wires and plates, to the solubility of the gases evolved, especially at the positive electrode.

When the positive and negative electrodes are equal in surface, the bubbles which rise from them in dilute sulphuric acid are always different in character. Those from the positive plate are exceedingly small, and separate instantly from every part of the surface of the metal, in consequence of its perfect cleanliness; whilst in the liquid they give it a hazy appearance, from their number and minuteness; are easily carried down by currents; and therefore not only present far greater surface of contact with the liquid than larger bubbles would do, but are retained a much longer time in mixture with it. But the bubbles at the negative surface, though they constitute twice the volume of the gas at the positive electrode, are nevertheless very inferior in number. They do not rise so universally from every part of the surface, but seemed to be evolved at different points; and though so much larger, they appear to cling to the metal, separating with difficulty from it, and when separated, instantly rising to the top of the liquid. If, therefore, oxygen and hydrogen had equal solubility in, or powers of combining with, water under similar circumstances, still, under the present conditions, the oxygen would be far the most liable to solution ; but when to these is added its well-known power of forming a compound with water, it is no longer surprising that such a compound should be produced in small quantities at the positive electrode ; and indeed the bleaching power which some philosophers have observed in a solution at this electrode, when chlorine and similar bodies have been carefully excluded, is probably due to the formation there, in this manner, of oxy-water.

That more gas was collected from the wires than from the plates, I attribute to the circumstance, that, as equal quantities are evolved in equal times, the bubbles at the wires having been more rapidly produced, in relation to any part of the surface, must have been much larger ; have been therefore in contact with the fluid by a much smaller surface, and for a much shorter time than those at the plates; hence less solution and a greater amount collected.

There was also another effect produced, especially by the use of large electrodes, which was both a consequence and a proof of the solution of part of the gas evolved there. The collected gas, when examined, was found to contain small portions of nitrogen. This I attribute to the presence of air dissolved in the acid used for decomposition. It is a well-known fact that when bubbles of gas but slightly soluble in water or solutions pass through them, the portion of this gas which is dissolved displaces a portion of that previously in union with the liquid : and so, in the decompositions under consideration, as the oxygen dissolves, it displaces a part of the air, or at least of the nitrogen, previously united to the acid ; and this effect takes place *most extensively* with large plates, because the gas evolved at them is in the most favorable condition for solution.

With the intention of avoiding this solubility of gases as much as possible, I arranged the decomposing plates in a vertical position, that the bubbles might quickly escape upwards, and that the downward currents in the fluid should not meet ascending currents of gas. This precaution I found to assist greatly in producing constant results, and especially in experiments to be hereafter referred to, in which other liquids than dilute sulphuric acid—as, for instance, solution of potash—were used.

The irregularities of the indications of the measurer proposed, arising from the solubility just referred to, are but small, and may be very nearly corrected by comparing the results of two or three experiments. They may also be almost entirely avoided by selecting that solution which is found to favor them in the least degree; and still further by collecting the hydrogen only, and using that as the indicating gas ; for being much less soluble than oxygen, being evolved with twice the rapidity and in larger bubbles, it can be collected more perfectly and in greater purity.

From the foregoing and many other experiments, it results that *variation in the size of the electrodes causes no variation in the chemical action of a given quantity of electricity upon water.*

The next point in regard to which the principle of constant electrochemical action was tested, was *variation of intensity.* In the first place, the preceding experiments were repeated, using batteries of an *equal* number of plates, *strongly* and weakly *charged ;* but the results were alike. They were then repeated, using batteries sometimes containing forty, and at other times only five pairs of plates ; but the results were still the same. *Variations therefore in intensity,* caused by difference in the strength of charge, or in the number of alternations used, *produced no difference as to the equal action of large and small electrodes.*

Still these results did not prove that variation in the intensity of the current was not accompanied by a corresponding variation in the electrochemical effects, since the action at *all* the surfaces might have increased or diminished together. The deficiency in the evidence is, however, completely supplied by the former experiments on different-sized electrodes ; for with variation in the size of these, a variation in the intensity must have occurred. The intensity of an electric current traversing conductors alike in their nature, quality, and length, is probably as the quantity of electricity passing through a given sectional area perpendicular to the current, divided by the time ; and therefore when large plates were contrasted with wires separated by an equal length of the same decomposing conductor, whilst one current of electricity passed through both arrangements, that electricity must have been in a very different state, as to *tension,* between the plates and between the wires ; yet the chemical results were the same.

The difference in intensity, under the circumstances described, may be easily shown practically, by arranging two decomposing apparatus as in Fig. 8, where the same fluid is subjected to the decomposing power of the same current of electricity, passing in the vessel A between large platina plates, and in the vessel B between small wires. If a third decomposing apparatus, such as that delineated in Fig. 7, be connected with the wires at ab, Fig. 8, it will serve sufficiently well, by the degree of decomposition

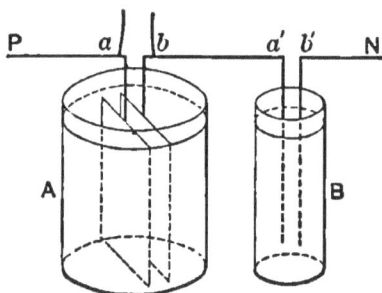

Fig. 8

occurring in it, to indicate the relative state of the two plates as to intensity; and if it then be applied in the same way, as a test of the state of the wires at $a'b'$, it will, by the increase of decomposition within, show how much greater the intensity is there than at the former points. The connections of P and N with the voltaic battery are of course to be continued during the whole time.

A third form of experiment in which difference of intensity was obtained, for the purpose of testing the principle of equal chemical action, was to arrange three volta-electrometers, so that after the electric current had passed through one, it should divide into two parts, each of which should traverse one of the remaining instruments, and should then reunite. The sum of the decomposition in the two latter vessels was always equal to the decomposition in the former vessel. But the *intensity* of the divided current could not be the same as that it had in its original state; and therefore *variation of intensity has no influence on the results if the quantity of electricity remain the same.* The experiment, in fact, resolves itself simply into an increase in the size of the electrodes.

The *third point*, in respect to which the principle of equal electrochemical action on water was tested, was *variation of the strength of the solution used.* In order to render the water a conductor, sulphuric acid had been added to it; and it did not seem unlikely that this substance, with many others, might render the water more subject to decomposition, the electricity remaining the same in quantity. But such did not prove to be the case. Diluted sulphuric acid, of different strengths, was introduced into different decomposing apparatus, and submitted simultaneously to the action of the same electric current. Slight differences occurred, as before, sometimes in one direction, sometimes in another ; but the final result was, that *exactly the same quantity of water was decomposed in all the solutions by the same quantity of electricity*, though the sulphuric acid in some was seventy-fold what it was in others. The strengths used were of specific gravity 1.495, and downwards.

When an acid having a specific gravity of about 1.336 was employed, the results were most uniform, and the oxygen and hydrogen most constantly in the right proportion to each other. Such an acid gave more gas than one much weaker acted upon by the same current, apparently because it had less solvent power. If the acid were very strong, then a remarkable disappearance of oxygen took place ; thus, one made by mixing two measures of strong oil of vitriol with one of water, gave forty-two volumes of hydrogen, but only twelve of oxygen. The hydrogen was very nearly the same with that evolved from acid of the specific gravity 1.232. I have not yet had time to examine minutely the circumstances attending the disappearance of the oxygen in this case, but imagine it is due to the formation of oxy-water, which Thénard has shown is favored by the presence of acid.

Although not necessary for the practical use of the instrument I am describing, yet as connected with the important point of constant electrochemical action upon water, I now investigated the effects produced by an electric current passing through aqueous solutions of acids, salts, and compounds, exceedingly different from each other in their nature, and found them to yield astonishingly uniform results. But many of them which are connected with a secondary action will be more usefully described hereafter.

When solutions of caustic potassa or soda, or sulphate of magnesia, or sulphate of soda, were acted upon by the electric current, just as much oxygen and hydrogen was evolved from them as from the diluted sulphuric acid, with which they were compared. When a solution of ammonia, rendered a better conductor by sulphate of ammonia, or a solution of subcarbonate of potassa, was experimented with, the *hydrogen* evolved was in the same quantity as that set free from the diluted sulphuric acid with which they were compared. Hence *changes in the nature of the solution do not alter the constancy of electrolytic action upon water.*

I have already said, respecting large and small electrodes, that change of order caused no change in the general effect. The same was the case with different solutions, or with different intensities ; and however the circumstances of an experiment might be varied, the results came forth exceedingly consistent, and proved that the electrochemical action was still the same.

I consider the foregoing investigation as sufficient to prove the very extraordinary and important principle with respect to WATER, *that when subjected to the influence of the electric current, a quantity of it is decomposed exactly proportionate to the quantity of electricity which has passed,* notwithstanding the thousand variations in the conditions and circumstances under which it may at the time be placed ; and further, that when the interference of certain secondary effects, together with the solution or recombination of the gas and the evolution of air, are guarded against, *the products of the decomposition may be collected with such accuracy as to afford a very excellent and valuable measurer of the electricity concerned in their evolution.*

The forms of instrument which I have given (Figs. 5, 6, 7) are probably those which will be found most useful, as they indicate the quantity of electricity by the largest volume of gases, and cause the least obstruction to the passage of the current. The fluid which my present experience leads me to prefer is a solution of sulphuric acid of specific gravity about 1.336, or from that to 1.25 ; but it is very essential that there should be no organic substance, nor any vegetable acid, nor other body, which, by being liable to the action of the oxygen or hydrogen evolved at the electrodes, shall diminish their quantity, or add other gases to them.

In many cases when the instrument is used as a *comparative standard*, or even as a *measurer*, it may be desirable to collect the hydrogen only, as being less liable to absorption or disappearance in other ways than the oxygen; whilst at the same time its volume is so large as to render it a good and sensible indicator. In such cases the first and second form of apparatus have been used (Figs. 3, 4). The indications obtained were very constant, the variations being much smaller than in those forms of apparatus collecting both gases ; and they can also be procured when solutions are used in comparative experiments, which, yielding no oxygen or only secondary results of its action, can give no indications if the educts at both electrodes be collected. Such is the case when solutions of ammonia, muriatic acid, chlorides, iodides, acetates, or other vegetable salts, etc., are employed.

In a few cases, as where solutions of metallic salts liable to reduction at the negative electrode are acted upon, the oxygen may be advantageously used as the measuring substance. This is the case, for instance, with sulphate of copper.

There are therefore two general forms of the instrument which I submit as a measurer of electricity: one, in which both the gases of the water decomposed are collected ; and the other, in which a single gas, as the hydrogen only, is used. When referred to as a *comparative instrument* (a use I shall now make of it very extensively), it will not often require particular precaution in the observation ; but when used as an *absolute measurer*, it will be needful that the barometric pressure and the temperature be taken into account, and that the graduation of the instruments should be to one scale : the hundredths and smaller divisions of a cubical inch are quite fit for this purpose, and the hundredth may be very conveniently taken as indicating a DEGREE of electricity.

It can scarcely be needful to point out further than has been done how this instrument is to be used. It is to be introduced into the course of the electric current, the action of which is to be exerted anywhere else, and if 60° or 70° of electricity are to be measured out, either in one or several portions, the current, whether strong or weak, is to be continued until the gas in the tube occupies that number of divisions or hundredths of a cubical inch. Or if a quantity competent to produce a certain effect is to be measured, the effect is to be obtained, and then the indication read off. In exact experiment it is necessary to correct the volume of gas for changes in temperature and pressure, and especially for moisture.* For

* For a simple table of correction for moisture, I may take the liberty of referring to my *Chemical Manipulation*, edition of 1830, p. 376.

the latter object the volta-electrometer (Fig. 7) is most accurate, as its gas can be measured over water, whilst the others retain it over acid or saline solutions.

I have not hesitated to apply the [term *degree* in analogy with the use made of it with respect to another most important imponderable agent—namely, heat ; and as the definite expansion of air, water, mercury, etc., is there made use of to measure heat, so the equally definite evolution of gases is here turned to a similar use for electricity.

The instrument offers the only *actual measurer* of voltaic electricity which we at present possess. For without being at all affected by variations in time or intensity, or alterations in the current itself, of any kind, or from any cause, or even of intermissions of action, it takes note with accuracy of the quantity of electricity which has passed through it, and reveals that quantity by inspection ; I have therefore named it a VOLTA-ELECTROMETER.

Another mode of measuring volta-electricity may be adopted with advantage in many cases, dependent on the quantities of metals or other substances evolved either as primary or as secondary results ; but I refrain from enlarging on this use of the products, until the principles on which their constancy depends have been fully established.

By the aid of this instrument I have been able to establish the definite character of electrochemical action in its most general sense; and I am persuaded it will become of the utmost use in the extensions of science which these views afford. I do not pretend to have made its detail perfect, but to have demonstrated the truth of the principle, and the utility of the application.*

[*Section VI., including thirteen pages, " On the Primary and Secondary Character of the Bodies Evolved at the Electrodes," is here omitted.*]

ON THE DEFINITE NATURE AND EXTENT OF ELECTROCHEMICAL
DECOMPOSITION

In the third series of the Researches, after proving the

* As early as the year 1811, Messrs. Gay-Lussac and Thénard employed chemical decomposition as a measurer of the electricity of the voltaic pile. See *Recherches Physico-chymiques*, p. 12. The principles and precautions by which it becomes an exact measure were of course not then known. *December*, 1838.

identity of electricities derived from different sources, and showing, by actual measurement, the extraordinary quantity of electricity evolved by a very feeble voltaic arrangement, I announced a law, derived from experiment, which seemed to me of the utmost importance to the science of electricity in general, and that branch of it denominated electrochemistry in particular. The law was expressed thus:* *The chemical power of a current of electricity is in direct proportion to the absolute quantity of electricity which passes.*

In the further progress of the successive investigations, I have had frequent occasion to refer to the same law, sometimes in circumstances offering powerful corroboration of its truth; and the present series already supplies numerous new cases in which it holds good. It is now my object to consider this great principle more closely, and to develop some of the consequences to which it leads. That the evidence for it may be more distinct and applicable, I shall quote cases of decomposition subject to as few interferences from secondary results as possible, effected upon bodies very simple, yet very definite in their nature.

In the first place, I consider the law as so fully established with respect to the decomposition of *water*, and under so many circumstances which might be supposed, if anything could, to exert an influence over it, that I may be excused entering into further detail respecting that substance, or even summing up the results here. I refer, therefore, to the whole of the subdivision of this series of Researches which contains the account of the *volta-electrometer*.

In the next place, I also consider the law as established with respect to *muriatic acid* by the experiments and reasoning already advanced, when speaking of that substance, in the subdivision respecting primary and secondary results.

I consider the law as established also with regard to *hydriodic acid* by the experiments and considerations already advanced in the preceding division of this series of Researches.

Without speaking with the same confidence, yet from the experiments described, and many others not described, relating to hydro-fluoric, hydro-cyanic, ferro-cyanic, and sulpho-cyanic acids, and from the close analogy which holds between these bodies and the hydracids of chlorine, iodine, bromine, etc., I

* [*See page* 7.]

consider these also as coming under subjection to the law, and assisting to prove its truth.

In the preceding cases, except the first, the water is believed to be inactive; but to avoid any ambiguity arising from its presence, I sought for substances from which it should be absent altogether; and, taking advantage of the law of conduction* already developed, I soon found abundance, among which *protochloride of tin* was first subjected to decomposition in the following manner : A piece of platina wire had one extremity coiled up into a small knob, and having been carefully weighed, was sealed hermetically into a piece of bottle-glass tube, so that the knob should be at the bottom of the tube within (Fig. 9). The tube was suspended by a piece of platina wire, so that the heat of a spirit-lamp could be applied to it. Recently fused protochloride of tin was introduced in sufficient quantity to occupy, when melted, about one-half of the tube ; the wire of the tube was connected with a volta-electrometer, which was itself connected  *Fig.* 9 with the negative end of a voltaic battery; and a platina wire connected with the positive end of the same battery was dipped into the fused chloride in the tube; being, however, so bent that it could not by any shake of the hand or apparatus touch the negative electrode at the bottom of the vessel. The whole arrangement is delineated in Fig. 10.

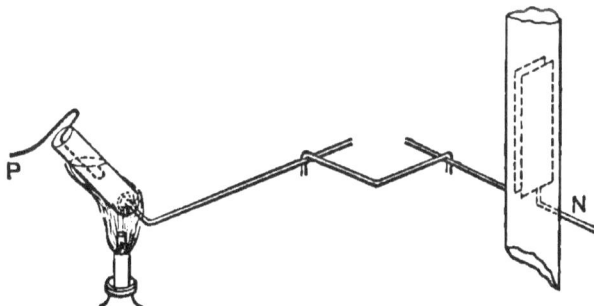

Fig. 10

Under these circumstances the chloride of tin was decomposed : the chlorine evolved at the positive electrode formed

* [*The law referred to asserts "the general assumption of conducting power by bodies as soon as they pass from the solid to the liquid state."*]

bichloride of tin, which passed away in fumes, and the tin evolved at the negative electrode combined with the platina, forming an alloy, fusible at the temperature to which the tube was subjected, and therefore never occasioning metallic communication through the decomposing chloride. When the experiment had been continued so long as to yield a reasonable quantity of gas in the volta-electrometer, the battery connection was broken, the positive electrode removed, and the tube and remaining chloride allowed to cool. When cold, the tube was broken open, the rest of the chloride and the glass being easily separable from the platina wire and its button of alloy. The latter when washed was then reweighed, and the increase gave the weight of the tin reduced.

I will give the particular results of one experiment, in illustration of the mode adopted in this and others, the results of which I shall have occasion to quote. The negative electrode weighed at first 20 grains; after the experiment it, with its button of alloy, weighed 23.2 grains. The tin evolved by the electric current at the *cathode* weighed therefore 3.2 grains. The quantity of oxygen and hydrogen collected in the volta-electrometer$=3.85$ cubic inches. As 100 cubic inches of oxygen and hydrogen, in the proportions to form water, may be considered as weighing 12.92 grains, the 3.85 cubic inches would weigh 0.49742 of a grain; that being, therefore, the weight of water decomposed by the same electric current as was able to decompose such weight of protochloride of tin as could yield 3.2 grains of metal. Now $0.49742 : 3.2 :: 9$ the equivalent of water is to 57.9, which should therefore be the equivalent of tin, if the experiment had been made without error, and if the electrochemical decomposition *is in this case also definite*. In some chemical works 58 is given as the chemical equivalent of tin, in others 57.9. Both are so near to the result of the experiment, and the experiment itself is so subject to slight causes of variation (as from the absorption of gas in the volta-electrometer), that the numbers leave little doubt of the applicability of the *law of definite action* in this and all similar cases of electro-decomposition.

It is not often I have obtained an accordance in numbers so near as I have just quoted. Four experiments were made on the protochloride of tin; the quantities of gas evolved in the volta-electrometer being from 2.05 to 10.29 cubic inches. The average of the four experiments gave 58.53 as the electrochemical equivalent for tin.

The chloride remaining after the experiment was pure protochloride of tin ; and no one can doubt for a moment that the equivalent of chlorine had been evolved at the *anode*, and, having formed bichloride of tin as a secondary result, had passed away.

Chloride of lead was experimented upon in a manner exactly similar, except that a change was made in the nature of the positive electrode ; for as the chlorine evolved at the *anode* forms no perchloride of lead, but acts directly upon the platina, it produces, if that metal be used, a solution of chloride of platina in the chloride of lead ; in consequence of which a portion of platina can pass to the *cathode,* and would then produce a vitiated result. I therefore sought for, and found in plumbago, another substance which could be used safely as the positive electrode in such bodies as chlorides, iodides, etc. The chlorine or iodine does not act upon it, but is evolved in the free state ; and the plumbago has no reaction, under the circumstances, upon the fused chloride or iodide in which it is plunged. Even if a few particles of plumbago should separate by the heat or the mechanical action of the evolved gas, they can do no harm in the chloride.

The mean of the three experiments gave the number of 100.85 as the equivalent for lead. The chemical equivalent is 103.5. The deficiency in my experiments I attribute to the solution of part of the gas in the volta-electrometer ; but the results leave no doubt on my mind that both the lead and the chlorine are, in this case, evolved in *definite quantities* by the action of a given quantity of electricity.

Chloride of antimony.—It was in endeavoring to obtain the electrochemical equivalent of antimony from the chloride, that I found reasons for the statement I have made respecting the presence of water in it in an earlier part of these Researches.

I endeavored to experiment upon the *oxide of lead* obtained by fusion and ignition of the nitrate in a platina crucible, but found great difficulty from the high temperature required for perfect fusion, and the powerful fluxing qualities of the substance. Green-glass tubes repeatedly failed. I at last fused the oxide in a small porcelain crucible, heated fully in a charcoal fire ; and, as it was essential that the evolution of the lead at the *cathode* should take place beneath the surface, the negative electrode was guarded by a green - glass tube, fused around it in such a manner as to expose only the knob of platina at the lower end (Fig. 11), so that it could be plunged beneath the surface, and thus exclude contact of air or oxygen with the lead reduced there. A platina wire was employed for the positive electrode, that metal not being subject to any action from the oxygen evolved against it. The arrangement is given in Fig. 12.

Fig. 11

In an experiment of this kind the equivalent for the lead came out 93.17, which is very much too small. This, I believe, was because of the small interval between the positive and negative electrodes in the oxide of lead ; so that it was not unlikely that some of the froth and bubbles formed by the oxygen at the *anode* should occasionally even touch the lead reduced at the *cathode,* and re-oxidize it. When I endeavored to correct this by having more litharge, the greater heat required to keep it all fluid caused a quicker action on the crucible, which was soon eaten through and the experiment stopped.

In one experiment of this kind I used borate of lead. It

Fig. 12

evolves lead, under the influence of the electric current, at the *anode,* and oxygen at the *cathode ;* and as the boracic acid is not either directly or incidentally decomposed during the operation, I expected a result dependent on the oxide of lead. The borate is not so violent a flux as the oxide, but it requires a higher temperature to make it quite liquid ; and if not very hot, the bubbles of oxygen cling to the positive electrode, and retard the transfer of electricity. The number for lead came out 101.29, which is so near to 103.5 as to show that the action of the current had been definite.

Oxide of bismuth.—I found this substance required too high

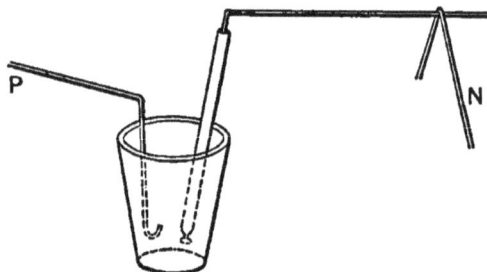

a temperature, and acted too powerfully as a flux, to allow of any experiment being made on it, without the application of more time and care than I could give at present.

The ordinary *protoxide of antimony*, which consists of one proportional of metal and one and a half of oxygen, was subjected to the action of the electric current in a green-glass tube, surrounded by a jacket of platina foil, and heated in a charcoal fire. The decomposition began and proceeded very well at first, apparently indicating, according to the general law, that this substance was one containing such elements and in such proportions as made it amenable to the power of the electric current. This effect I have already given reasons for supposing may be due to the presence of a true protoxide, consisting of single proportionals. The action soon diminished, and finally ceased, because of the formation of a higher oxide of the metal at the positive electrode. This compound, which was probably the peroxide, being infusible and insoluble in the protoxide, formed a crystalline crust around the positive electrode ; and thus insulating it, prevented the transmission of the electricity. Whether, if it had been fusible and still immiscible, it would have decomposed, is doubtful, because of its departure from the required composition. It was a very natural secondary product at the positive electrode. On opening the tube it was found that a little antimony had been separated at the negative electrode ; but the quantity was too small to allow of any quantitative result being obtained.

Iodide of lead.—This substance can be experimented with in tubes heated by a spirit-lamp ; but I obtained no good results from it, whether I used positive electrodes of platina or plumbago. In two experiments the numbers for the lead came out only 75.46 and 73.45, instead of 103.5. This I attribute to the formation of a periodide at the positive electrode, which, dissolving in the mass of liquid iodide, came in contact with the lead evolved at the negative electrode, and dissolved part of it, becoming itself again protiodide. Such a periodide does exist ; and it is very rarely that the idiode of lead formed by precipitation, and well washed, can be fused without evolving much iodine from the presence of this percompound ; nor does crystallization from its hot aqueous solution free it from this substance. Even when a little of the protiodide and iodine are merely rubbed together in a mortar a portion of the periodide is formed. And though it is decomposed by being fused and heated to dull redness for a few minutes, and the whole reduced to protiodide, yet that is not at all opposed to the possibility, that a little of that which is formed in great excess of iodine at the *anode*, should be carried by the rapid currents in the liquid into contact with the *cathode*.

This view of the result was strengthened by a third experiment, where the space between the electrodes was increased to one-third of an inch; for now the interfering effects were much diminished, and the number of the lead came out 89.04; and it was fully confirmed by the results obtained in the cases of *transfer* to be immediately described.

The experiments on iodide of lead, therefore, offer no exception to the *general law* under consideration, but on the contrary may, from general considerations, be admitted as included in it.

Protiodide of tin. — This substance, when fused,.conducts and is decomposed by the electric current, tin is evolved at the *anode*, and periodide of tin as a secondary result at the *cathode*. The temperature required for its fusion is too high to allow of the production of any results for weighing.

Iodide of potassium was subjected to electrolytic action in a tube, like that in Fig. 9. The negative electrode was a globule of lead, and I hoped in this way to retain the potassium, and obtain results that could be weighed and compared with the volta-electrometer indication; but the difficulties dependent upon the high temperature required, the action upon the glass, the fusibility of the platina induced by the presence of the lead, and other circumstances, prevented me from procuring such results. The iodide was decomposed with the evolution of iodine at the *anode*, and of potassium at the *cathode*, as in former cases.

In some of these experiments several substances were placed in succession, and decomposed simultaneously by the same electric current; thus, protochloride of tin, chloride of lead, and water, were thus acted on at once. It is needless to say that the results were comparable, the tin, lead, chlorine, oxygen, and hydrogen evolved being *definite in quantity* and electrochemical equivalents to each other.

Let us turn to another kind of proof of the *definite chemical action of electricity*. If any circumstances could be supposed to exert an influence over the quantity of the matters evolved during electrolytic action, one would expect them to be present when electrodes of different substances, and possessing very different chemical affinities for such matters, were used. Platina has no power in dilute sulphuric acid of combining with the oxygen at the *anode,* though the latter be evolved in the nascent state against it. Copper, on the other hand, immediately unites with the oxygen, as the electric current sets it free from the hydrogen ; and zinc is not only able to combine with it, but can, without any help from the electricity, abstract it directly from the water, at the same time setting torrents of hydrogen free. Yet in cases where these three substances were used as the positive electrodes, in three similar portions of the same dilute sulphuric acid, specific gravity 1.336, precisely the same quantity of water was decomposed by the electric current, and precisely the same quantity of hydrogen set free at the *cathodes* of the three solutions.

The experiment was made thus : Portions of the dilute sulphuric acid were put into three basins. Three volta-electrometer tubes, of the form Figs. 1, 3, were filled with the same acid, and one inverted in each basin. A zinc plate, connected with the positive end of a voltaic battery, was dipped into the first basin, forming the positive electrode there, the hydrogen, which was abundantly evolved from it by the direct action of the acid, being allowed to escape. A copper plate, which dipped into the acid of the second basin, was connected with the negative electrode of the *first* basin ; and a platina plate, which dipped into the acid of the third basin, was connected with the negative electrode of the *second* basin. The negative electrode of the third basin was connected with a volta-electrometer, and that with the negative end of the voltaic battery.

Immediately that the circuit was complete, the *electrochemical action* commenced in all the vessels. The hydrogen still rose in apparently undiminished quantities from the positive zinc electrode in the first basin. No oxygen was evolved at the positive copper electrode in the second basin, but a sulphate of copper was formed there ; whilst in the third basin the positive platina electrode evolved pure oxygen gas, and was itself unaffected. But in *all* the basins the hydrogen liberated at the *negative* platina electrodes was the *same in quantity,* and the same with the volume of hydrogen evolved in the volta-elec-

c   33

trometer, showing that in all the vessels the current had decomposed an equal quantity of water. In this trying case, therefore, the *chemical action of electricity* proved to be *perfectly definite.*

A similar experiment was made with muriatic acid diluted with its bulk of water. The three positive electrodes were zinc, silver, and platina; the first being able to separate and combine with the chlorine *without* the aid of the current; the second combining with the chlorine only after the current had set it free; and the third rejecting almost the whole of it. The three negative electrodes were, as before, platina plates fixed within glass tubes. In this experiment, as in the former, the quantity of hydrogen evolved at the *cathodes* was the same for all, and the same as the hydrogen evolved in the volta-electrometer. I have already given my reasons for believing that in these experiments it is the muriatic acid which is directly decomposed by the electricity; and the results prove that the quantities so decomposed are *perfectly definite* and proportionate to the quantity of electricity which has passed.

In this experiment the chloride of silver formed in the second basin retarded the passage of the current of electricity, by virtue of the law of conduction before described,* so that it had to be cleaned off four or five times during the course of the experiment; but this caused no difference between the results of that vessel and the others.

Charcoal was used as the positive electrode in both sulphuric and muriatic acids; but this change produced no variation of the results. A zinc positive electrode, in sulphate of soda or solution of common salt, gave the same constancy of operation.

Experiments of a similar kind were then made with bodies altogether in a different state—*i. e.*, with *fused* chlorides, iodides, etc. I have already described an experiment with fused chloride of silver, in which the electrodes were of metallic silver, the one rendered negative becoming increased and lengthened by the addition of metal, whilst the other was dissolved and eaten away by its abstraction. This experiment was repeated, two weighed pieces of silver wire being used as the electrodes, and a volta-electrometer included in the circuit. Great care was taken to withdraw the negative electrode so regularly

*[*See foot-note*, p. 27.]

and steadily that the crystals of reduced silver should not form a *metallic* communication beneath the surface of the fused chloride. On concluding the experiment the positive electrode was reweighed, and its loss ascertained. The mixture of chloride of silver and metal, withdrawn in successive portions at the negative electrode, was digested in solution of ammonia, to remove the chloride, and the metallic silver remaining also weighed : it was the reduction at the *cathode*, and exactly equalled the solution at the *anode;* and each portion was as nearly as possible the equivalent to the water decomposed in the volta-electrometer.

The infusible condition of the silver at the temperature used, and the length and ramifying character of its crystals, render the above experiment difficult to perform, and uncertain in its results. I therefore wrought with chloride of lead, using a green-glass tube, formed as in Fig. 13. A weighed platina wire was fused into the bottom of a small tube, as before described. The tube was then bent to an angle, at about half an inch

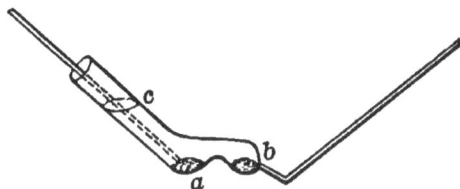

Fig. 13

distance from the closed end ; and the part between the angle and the extremity being softened, was forced upward, as in the figure, so as to form a bridge, or rather separation, producing two little depressions or basins, *a, b*, within the tube. This arrangement was suspended by a platina wire, as before, so that the heat of a spirit-lamp could be applied to it, such inclination being given to it as would allow all air to escape during the fusion of the chloride of lead. A positive electrode was then provided, by bending up the end of a platina wire into a knot, and fusing about twenty grains of metallic lead on to it, in a small closed tube of glass, which was afterwards broken away. Being so furnished, the wire with its lead was weighed, and the weight recorded.

Chloride of lead was now introduced into the tube, and carefully fused. The leaded electrode was also introduced, after which the metal at its extremity soon melted. In this state of things the tube was filled up to *c* with melted chloride of lead ; the end of the electrode to be rendered negative was in the basin b, and, the electrode of melted lead was retained in the basin a, and, by connection with the proper conducting wire of a voltaic battery, was rendered positive. A volta-electrometer was included in the circuit.

Immediately upon the completion of the communication with the voltaic battery, the current passed, and decomposition proceeded. No chlorine was evolved at the positive electrode ; but as the fused chlorine was transparent, a button of alloy could be observed gradually forming and increasing in size at b, whilst the lead at a could also be seen gradually to diminish. After a time the experiment was stopped, the tube allowed to cool, and broken open ; the wires, with their buttons, cleaned and weighed ; and their change in weight compared with the indication of the volta-electrometer.

In this experiment the positive electrode had lost just as much lead as the negative one had gained, and the loss and gain were very nearly the equivalents of the water decomposed in the volta-electrometer, giving for lead the number 101.5. It is therefore evident, in this instance, that causing a *strong affinity* or *no affinity,* for the substance evolved at the *anode*, to be active during the experiment, produces no variation in the definite action of the electric current.

A similar experiment was then made with iodide of lead, and in this manner all confusion from the formation of a periodide avoided. No iodine was evolved during the whole action, and finally the loss of lead at the *anode* was the same as the gain at the *cathode*, the equivalent number by comparison with the result in the volta-electrometer being 103.5.

Then protochloride of tin was subjected to the electric current in the same manner, using, of course, a tin positive electrode. No bichloride of tin was now formed. On examining the two electrodes, the positive had lost precisely as much as the negative had gained ; and by comparison with the volta-electrometer, the number for tin came out 59.

It is quite necessary in these and similar experiments to examine the interior of the bulbs of alloy at the ends of the conducting wires ; for occasionally, and especially with those which have been positive, they are cavernous, and contain portions of the chloride or iodide used, which must be removed before the final weight is ascertained. This is more usually the case with lead than tin.

All these facts combine into, I think, an irresistible mass of evidence, proving the truth of the important proposition which I at first laid down—namely, *that the chemical power of a current of electricity is in direct proportion to the absolute quantity of electricity which passes.* They prove, too, that this is not merely true of one substance, as water, but generally with all electrolytic bodies ; and, further, that the results obtained with any *one substance* do not merely agree amongst themselves, but also with those obtained from *other substances,* the whole combining together into *one series of definite electrochemical actions.* I do not mean to say that no exceptions will appear ; perhaps some may arise, especially amongst substances existing only by weak affinity ; but I do not expect that any will seriously disturb the result announced. If, in the well considered, well examined, and, I may surely say, well-ascertained doctrines of the definite nature of ordinary chemical affinity, such exceptions occur, as they do in abundance, yet, without being allowed to disturb our minds as to the general conclusion, they ought also to be allowed if they should present themselves at this, the opening of a new view of electrochemical action ; not being held up as obstructions to those who may be engaged in rendering that view more and more perfect, but laid aside for a while, in hopes that their perfect and consistent explanation will ultimately appear.

The doctrine of *definite electrochemical action* just laid down, and, I believe, established, leads to some new views of the relations and classifications of bodies associated with or subject to this action. Some of these I shall proceed to consider.

In the first place, compound bodies may be separated into two great classes—namely, those which are decomposable by the electric current, and those which are not : of the latter, some are conductors, others non-conductors, of voltaic electricity.* The former do not depend for their decomposability upon the nature of their elements only ; for, of the same two elements, bodies may be formed, of which one shall belong to one class and another to the other class; but probably on the proportions also. It is further remarkable, that with very few, if any, exceptions, these decomposable bodies are exactly

---

* I mean here, by voltaic electricity, merely electricity from a most abundant source, but having very small intensity.

those governed by the remarkable law of conduction I have before described; for that law does not extend to the many compound fusible substances that are excluded from this class. I propose to call bodies of this, the decomposable class, *Electrolytes*.

Then, again, the substances into which these divide, under the influence of the electric current, form an exceedingly important general class. They are combining bodies; are directly associated with the fundamental parts of the doctrine of chemical affinity; and have each a definite proportion, in which they are always evolved during electrolytic action. I have proposed to call these bodies generally *ions*, or particularly *anions* and *cations*, according as they appear at the *anode* or *cathode;* and the numbers representing the proportions in which they are evolved *electrochemical equivalents.* Thus hydrogen, oxygen, chlorine, iodine, lead, tin, are *ions*; the three former* are *anions*, the two metals are *cations*, and 1, 8, 36, 125, 104, 58, are their *electrochemical equivalents* nearly.

A summary of certain points already ascertained respecting *electrolytes, ions,* and *electrochemical equivalents* may be given in the following general form of propositions, without, I hope, including any serious error.

I. A single *ion, i. e.,* one not in combination with another, will have no tendency to pass to either of the electrodes, and will be perfectly indifferent to the passing current, unless it be itself a compound of more elementary *ions,* and so subject to actual decomposition. Upon this fact is founded much of the proof adduced in favor of the new theory of electrochemical decomposition, which I put forth in a former series of these Researches.

II. If one *ion* be combined in right proportions with another strongly opposed to it in its ordinary chemical relations, *i. e.,* if an *anion* be combined with a *cation,* then both will travel, the one to the *anode,* the other to the *cathode,* of the decomposing body.

III. If, therefore, an *ion* pass towards one of the electrodes, another *ion* must also be passing simultaneously to the other electrode, although, from secondary action, it may not make its appearance.

IV. A body decomposable directly by the electric current,

---

* [*Oxygen, chlorine, and iodine are undoubtedly here referred to.*]

i. e., an *electrolyte,* must consist of two *ions,* and must also render them up during the act of decomposition.

V. There is but one *electrolyte* composed of the same two elementary *ions ;* at least such appears to be the fact, dependent upon a law, that *only single electrochemical equivalents of elementary ions can go to the electrodes, and not multiples.*

VI. A body not decomposable when alone, as boracic acid, is not directly decomposable by the electric current when in combination. It may act as an *ion* going wholly to the *anode* or *cathode,* but does not yield up its elements, except occasionally by a secondary action. Perhaps it is superfluous for me to point out that this proposition has *no relation* to such cases as that of water, which, by the presence of other bodies, is rendered a better conductor of electricity, and *therefore* is more freely decomposed.

VII. The nature of the substance of which the electrode is formed, provided it be a conductor, causes no difference in the electro-decomposition, either in kind or degree ; but it seriously influences, by secondary action, the state in which the *ions* finally appear. Advantage may be taken of this principle in combining and collecting such *ions* as, if evolved in their free state, would be unmanageable.*

VIII. A substance which, being used as the electrode, can combine with the *ion* evolved against it, is also, I believe, an *ion,* and combines, in such cases, in the quantity represented by its *electrochemical equivalent.* All the experiments I have made agree with this view ; and it seems to me, at present, to result as a necessary consequence. Whether, in the secondary actions that take place, where the *ion* acts not upon the matter of the electrode, but on that which is around it in the liquid, the same consequence follows, will require more extended investigation to determine.

IX. Compound *ions* are not necessarily composed of electrochemical equivalents of simple *ions.* For instance, sulphuric

---

* It will often happen that the electrodes used may be of such a nature as, with the fluid in which they are immersed, to produce an electric current, either according with or opposing that of the voltaic arrangement used, and in this way, or by direct chemical action, may sadly disturb the results. Still, in the midst of all these confusing effects, the electric current, which actually passes in any direction through the body suffering decomposition, will produce its own definite electrolytic action.

acid, boracic acid, phosphoric acid, are *ions*, but not *electrolytes*, *i. e.*, not composed of electrochemical equivalents of simple *ions*.

X. Electrochemical equivalents are always consistent, *i. e.*, the same number which represents the equivalent of a substance, *A*, when it is separating from a substance, *B*, will also represent *A* when separating from a third substance, *C*. Thus, 8 is the electrochemical equivalent of oxygen, whether separating from hydrogen, or tin, or lead; and 103.5 is the electrochemical equivalent of lead, whether separating from oxygen, or chlorine, or iodine.

XI. Electrochemical equivalents coincide, and are the same, with ordinary chemical equivalents.

By means of experiment and the preceding propositions, a knowledge of *ions* and their electrochemical equivalents may be obtained in various ways.

In the first place, they may be determined directly, as has been done with hydrogen, oxygen, lead, and tin, in the numerous experiments already quoted.

In the next place, from propositions II. and III., may be deduced the knowledge of many other *ions* and also their equivalents. When chloride of lead was decomposed, platina being used for both electrodes, there could remain no more doubt that chlorine was passing to the *anode*, although it combined with the platina there, than when the positive electrode, being of plumbago, allowed its evolution in the free state; neither could there, in either case, remain any doubt that for every 103.5 parts of lead evolved at the *cathode*, 36 parts of chlorine were evolved at the *anode*, for the remaining chloride of lead was unchanged. So also, when in a metallic solution, one volume of oxygen, or a secondary compound containing that proportion, appeared at the *anode*, no doubt could arise that hydrogen, equivalent to two volumes, had been determined to the *cathode*, although, by a secondary action, it had been employed in reducing oxides of lead, copper, or other metals, to the metallic state. In this manner, then, we learn from the experiments already described in these Researches that chlorine, iodine, bromine, fluorine, calcium, potassium, strontium, magnesium, manganese, etc., are *ions*, and that their *electrochemical equivalents* are the same as their *ordinary chemical equivalents*.

Propositions IV. and V. extend our means of gaining information. For if a body of known chemical composition is found to be decomposable, and the nature of the substance evolved as a primary or even a secondary result at one of the electrodes, be ascertained, the electrochemical equivalent of that body may be deduced from the known constant composition of the substance evolved. Thus, when fused protiodide of tin is decomposed by the voltaic current, the conclusion may be drawn that both the iodine and tin are *ions*, and that the proportions in which they combine in the fused compound express their electrochemical equivalents. Again, with respect to the fused iodide of potassium, it is an electrolyte; and the chemical equivalents will also be the electrochemical equivalents.

If proposition VIII. sustain extensive experimental investigation, then it will not only help to confirm the results obtained by the use of the other propositions, but will give abundant original information of its own.

In many instances the *secondary results* obtained by the action of the evolved *ion* on the substances present in the surrounding liquid or solution will give the electrochemical equivalent. Thus, in the solution of acetate of lead, and, as far as I have gone, in other proto-salts subjected to the reducing action of the nascent hydrogen at the *cathode*, the metal precipitated has been in the same quantity as if it had been a primary product (provided no free hydrogen escaped there), and therefore gave accurately the number representing its electrochemical equivalent.

Upon this principle it is that secondary results may occasionally be used as measurers of the volta-electric current; but there are not many metallic solutions that answer this purpose well; for unless the metal is easily precipitated hydrogen will be evolved at the *cathode* and vitiate the result. If a soluble peroxide is formed at the *anode*, or if the precipitated metal crystallize across the solution and touch the positive electrode, similar vitiated results are obtained. I expect to find in some salts, as the acetates of mercury and zinc, solutions favorable for this use.

After the first experimental investigations to establish the definite chemical action of electricity, I have not hesitated to apply the more strict results of chemical analysis to correct the numbers obtained as electrolytic results. This, it is evident, may be done in a great number of cases, without using too much liberty towards the due severity of scientific research.

The series of numbers representing electrochemical equivalents must, like those expressing the ordinary equivalents of chemically acting bodies, remain subject to the continual correction of experiment and sound reasoning. I give the following brief table of *ions* and their electrochemical equivalents, rather as a specimen of a first attempt than as anything that can supply the want, which must very quickly be felt, of a full and complete tabular account of this class of bodies. Looking forward to such a table as of extreme utility (if well constructed) in developing the intimate relation of ordinary chemical affinity to electrical actions, and identifying the two, not to the imagination merely, but to the conviction of the senses and a sound judgment, I may be allowed to express a hope, that the endeavor will always be to make it a table of *real*, and not *hypothetical*, electrochemical equivalents; for we shall else overrun the facts, and lose all sight and consciousness of the knowledge lying directly in our path.

The equivalent numbers do not profess to be exact, and are taken almost entirely from the chemical results of other philosophers in whom I could repose more confidence, as to these points, than in myself.

TABLE OF IONS

ANIONS

Oxygen	8	Phosphoric acid	35.7
Chlorine	35.5	Carbonic acid	22
Iodine	126	Boracic acid	24
Bromine	78.3	Acetic acid	51
Fluorine	18.7	Tartaric acid	66
Cyanogen	26	Citric acid	58
Sulphuric acid	40	Oxalic acid	36
Selenic acid	64	Sulphur (?)	16
Nitric acid	54	Selenium (?)	
Chloric acid	75.5	Sulpho-cyanogen	

CATIONS

Hydrogen	1	Calcium	20.5
Potassium	39.2	Magnesium	12.7
Sodium	23.3	Manganese	27.7
Lithium	10	Zinc	32.5
Barium	68.7	Tin	57.9
Strontium	43.8	Lead	103.5

CATIONS—*Continued*

Iron	28	Potassa	47.2
Copper	31.6	Soda	31.3
Cadmium	55.8	Lithia	18
Cerium	46	Baryta	76.7
Cobalt	29.5	Strontia	51.8
Nickel	29.5	Lime	28
Antimony	64.6 (?)	Magnesia	20.7
Bismuth	71	Alumina	(?)
Mercury	200	Protoxides generally	
Silver	108	Quinia	171.6
Platina	98.6 (?)	Cinchona	160
Gold	(?)	Morphia	290
		Vegeto-alkalies generally	
Ammonia	17		

This table might be further arranged into groups of such substances as either act with, or replace, each other. Thus, for instance, acids and bases act in relation to each other; but they do not act in association with oxygen, hydrogen, or elementary substances. There is indeed little or no doubt that, when the electrical relations of the particles of matter come to be closely examined, this division must be made. The simple substances, with cyanogen, sulpho-cyanogen, and one or two other compound bodies, will probably form the first group; and the acids and bases, with such analogous compounds as may be proved to be *ions*, the second group. Whether these will include all *ions*, or whether a third class of more complicated results will be required, must be decided by future experiments.

It is *probable* that all our elementary bodies are *ions*, but that is not yet certain. There are some, such as carbon, phosphorus, nitrogen, silicon, boron, aluminum, the right of which to the title of *ion* it is desirable to decide as soon as possible. There are also many compound bodies, and among them alumina and silica, which it is desirable to class immediately by unexceptional experiments. It is also *possible*, that all combinable bodies, compound as well as simple, may enter into the class of *ions;* but at present it does not seem to me probable. Still the experimental evidence I have is so small in proportion to what must gradually accumulate around, and bear upon, this point, that I am afraid to give a strong opinion upon it.

I think I cannot deceive myself in considering the doctrine of definite electrochemical action as of the utmost importance. It touches by its facts more directly and closely than any former fact, or set of facts, have done, upon the beautiful idea, that ordinary chemical affinity is a mere consequence of the electrical attractions of the particles of different kinds of matter; and it will probably lead us to the means by which we may enlighten that which is at present so obscure, and either fully demonstrate the truth of the idea or develop that which ought to replace it.

A very valuable use of electrochemical equivalents will be to decide, in cases of doubt, what is the true chemical equivalent, or definite proportional, or atomic number of a body; for I have such conviction that the power which governs electro-decomposition and ordinary chemical attractions is the same; and such confidence in the overruling influence of those natural laws which render the former definite, as to feel no hesitation in believing that the latter must submit to them also. Such being the case, I can have no doubt that, assuming hydrogen as 1, and dismissing small fractions for the simplicity of expression, the equivalent number or atomic weight of oxygen is 8, of chlorine 36, of bromine 78.4, of lead 103.5, of tin 59, etc., notwithstanding that a very high authority doubles several of these numbers.

ROYAL INSTITUTION, *December* 31, 1833.

MICHAEL FARADAY was born at Newington, a suburb of London, September 22, 1791. His parents were of humble origin, and their straitened circumstances prevented his obtaining more than the rudiments of an early education. When fourteen years old he was apprenticed to a bookbinder, and in this capacity was able to pick up considerable scientific knowledge himself from the books which came in for binding. He was, moreover, able to further his education by attending lectures on natural philosophy at the house of a Mr. Tatum, and also occasional lectures at the Royal Institution by Sir Humphry Davy. It was largely on account of the remarkable excellence of the notes which he worked up on these latter lectures, that in March, 1813, six months after he had completed his apprenticeship, Davy secured for him the appointment of assistant in chemistry at the Royal Institution. In October of the same year, he left the institution to accompany Davy, as secretary and chemical assistant, on an extended continental tour, which proved to be of the greatest value to him, not only for the broadening influence of travel it afforded, but also for the acquaintances he made with all the leading scientists in Europe, many of whom became later his friends and correspondents. On his return to England in the spring of 1815 he obtained a reappointment, and took up his residence at the Royal Institution. From this time he entered upon that wonderful career of investigation which soon placed him without a peer among his contemporaries.

In 1825 Faraday was appointed Director of the Laboratory at the Royal Institution. The following year he instituted the Friday Evening Lectures, and there demonstrated his wonderful ability to interest and hold the attention of popular audiences, by his personal magnetism and his clear and entertaining exposition of scientific questions. He was made a member of many learned societies, including the Royal Society, and received the degree of D.C.L. from the University of Oxford. He declined positions more remunerative than that at the Royal Institution, out of loyalty to the institution which had done so much for his advancement, and refused to devote his time to the commercial development of his many discoveries, although the pecuniary inducements were great. In 1840 he became an elder in the Sandemarian Church, of which he was an active member.

In 1821 Faraday married Miss Sarah Barnard, the marriage being an ideally happy one. He and his wife lived at the Royal Institution until 1858, when the Queen placed a house at Hampton Court at their disposal. Here he spent the remaining years of his life, continuing active work, however, at the Royal Institution until 1865.

He passed peacefully away at Hampton Court, August 25, 1867. His headstone, in Highgate Cemetery, bears the simple inscription :

MICHAEL FARADAY

Born 22nd September

1791

Died 25th August

1867

Faraday's scientific publications include no less than 163 papers (see Royal Society Catalogue) on the most diverse chemical and physical subjects. Among the most important of these, in addition to those on electrochemistry leading to the discovery of the law which bears his name, may be mentioned those on the liquefaction and solidification of gases, 1823 and 1845; on the discovery of benzine, 1825 ; on the seat of the electromotive force in the voltaic cell, 1834 ; on the laws of static and electromagnetic induction and the development of the concept of lines of force, 1831–1838 ; on the para and dia-magnetic properties of bodies, 1846 ; and, perhaps the greatest of all, on the discovery of the electromagnetic rotation of polarized light, 1845.

ON THE MIGRATION OF IONS DURING ELECTROLYSIS

BY

W. HITTORF

(Poggendorff's *Annalen*, **89,** 177, 1853 ; Ostwald's *Klassiker der Exakten Wissenschaften*, No. 21)

CONTENTS

ON THE MIGRATION OF IONS DURING ELECTROLYSIS

BY

W. HITTORF

THE explanation which we now give of the process of electrolysis was first proposed, in its general outlines, by Grotthuss in 1805. According to it, the two ions which are simultaneously set free do not come from the same molecule* of the electrolyte, but belong to different ones—namely, those which are in immediate contact with the electrodes. The other components of the compound from which they separate unite at once with the opposite components of the next adjacent molecules; this process takes place between the opposite components of all adjacent molecules in the interior of the electrolyte, and holds them together.

"I conclude from this," remarks Grotthuss,† "that were it possible to produce in water a galvanic current flowing in a circle, without the introduction of metallic conductors, all water particles lying in this circle would be decomposed and immediately after recombined; whence it follows that this water, although it actually undergoes galvanic decomposition in all of its particles, would nevertheless always remain water."

This conception of electrolysis was so natural that it could not fail to supersede the other more or less far-fetched hypotheses which assumed both liberated ions to arise from the same molecule of the electrolyte. It explained without further assumption, the numerous experiments which H. Davy‡ published soon afterwards on the transference of components to the electrodes.

---

* [*For greater clearness the term "atom," used throughout by Hittorf, has been translated "molecule" whenever the concept of molecule is intended.*]

† *Phys. Chem. Forsch.*, p. 123.

‡ *Gilb. Ann.*, **28**, 26.

D   49

MEMOIRS ON THE FUNDAMENTAL

The tardy appearance of the ions of an electrolyte which is not in direct contact with the poles, and their failure to appear at all when separated from the electrodes by a liquid with which they form an insoluble compound, were excellent proofs of the theory furnished by Davy.

Notwithstanding the clear conception of electrolysis which Grotthuss had up to this point, as indicated in the remark which I have given above in his own words (we easily realize to-day, as is well known, the premise of the conclusion by an induction current), he fell into serious error in attempting to further fathom the phenomenon. He conceived it to be produced as follows: the metals between which the electrolyte is placed are the seat of two forces, which vary inversely as the square of the distance, and which, acting oppositely on the two components, repel the one and attract the other. All physicists who turned their attention to this subject favored this view more or less for a long time ; the name of the poles which was given to the immersed metals corresponded to it. Grotthuss was, however, herein so far in advance of others that he considered (contrary to his hypothesis, to be sure) the forces acting on each particle of the electrolyte everywhere equal in the circle, an assumption which, as is known, is correct for the simplest conditions of the experiment.

Faraday was the first to penetrate deeper into the phenomenon. He conceived the cause of it in exactly the reverse manner, and was thereby led to the great discovery of the fundamental electrolytic action of the current, which now forms the basis of all investigations in electrolysis. By means of this change he brought the theory into harmony with Ohm's law, without knowing the latter.

"I conceive," he says, in § 524 of his *Experimental Researches,* * "the effects to arise from forces which are *internal,* relative to the matter under decomposition, and not *external,* as they might be considered, if directly dependent on the poles. I suppose that the effects are due to a modification, by the electric current, of the chemical affinity of the particles through or by which that current is passing, giving them the power of acting more forcibly in one direction than in another, and consequently making them travel by a series of successive decom-

* Pogg. *Ann.*, **32**, 435.

positions and recompositions in opposite directions, and finally causing their expulsion or exclusion at the boundaries of the body under decomposition, in the direction of the current, *and that* in larger or smaller quantities, according as the current is more or less powerful. I think, therefore, it would be more philosophical, and more directly expressive of the facts, to speak of such a body, in relation to the current passing through it, rather than to the poles, as they are usually called, in contact with it; and say that whilst under decomposition, oxygen, chlorine, iodine, acids, etc., are rendered at its negative extremity, and combustibles, metals, alkalies, bases, etc., at its· positive extremity.

" The poles, § 556,* are merely the surfaces or doors by which the electricity enters into or passes out of the substance suffering decomposition. They limit the extent of that substance in the course of the electric current, being its *terminations* in that direction; hence the elements evolved pass so far and no further."

In this way Faraday, for the first time, explains chemical decomposition with definiteness, as the conduction of the electric current through the electrolyte. / He proves the important relation that† " *the sum of chemical decomposition is constant for every section taken across a decomposing conductor, uniform in its nature, at whatever distance the poles may be from each other or from the section; . . . provided the current of electricity be retained in constant quantity. . .*"

Our conception even to-day of the process of electrolytic decomposition is embraced in these laws. In a later paper ‡ Faraday expressed the belief that they would need modification. The chemical theory of the galvanic cell, which he so energetically sought to defend, inclined him, primarily, to make this statement, as well as the fact that electrolytes often conduct weak currents without any decomposition being perceptible. Both points, however, have since been satisfactorily explained by science, without in any way affecting the postulated laws. On the contrary, every more exact investigation has only furnished a new confirmation of them.

We usually picture the process to ourselves by means of a row

---

* Pogg. *Ann.*, **32**, 450.
† Ibid., **32**, 426.
‡ Ibid., **35**, 259.

of adjacent molecules, as shown in Fig. 1. It is assumed in the figure that the distance between the neighboring molecules of the electrolyte is greater than that between the chemically bound ions of each individual molecule. This assumption is certainly permissible in those cases which we shall alone have to consider later — namely, those in which the electrolyte is brought into the liquid state by means of a solvent.

The first action of the current consists* in bringing the particles of the body to be decomposed into such a position that the cation of each molecule is turned towards the cathode, and

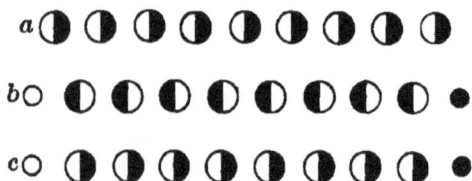

Fig. 1

the anion towards the anode. The two ions then separate from each other, move in opposite directions, and thereby meet with the neighboring ions likewise migrating (Fig. 1, *b*). By this process, however, they have arrived in a position where each anion is turned towards the cathode, and each cation towards the anode. There must therefore result a rotation of each molecule, and the reverse position be established, if the same constituent is to be continuously liberated at the same electrode (Fig. 1, *c*).

It would certainly be of great importance if we could represent these motions, to which the smallest particles of an electrolyte are subjected during the passage of the current, more definitely than in these most general outlines. They would not only throw light on the nature of electricity, but also on the chemical constitution of bodies.

In many cases it seems possible to determine by experiment the relative distances through which the two ions move during electrolysis. As we shall be concerned only with this point in what follows, we will give prominence to it alone in the figure.

* See *Faraday*, § 1705; Pogg. *Ann.; Ergänzungsband*, I., 263.

For this purpose let us adopt the method of representation given by Berzelius in his works, in which the two ions are represented one below the other, and supposed to move by each other in a horizontal direction (Fig. 2). It is assumed that the electrolyte is brought into the liquid state by means of an indifferent non-conducting solvent.

If we can divide the liquid at any definite place, we shall find that the ions in each portion are in a different proportion after electrolysis has taken place than before. This proportion is determined by the distance through which each ion moves during the passage of the current.

If, for example, we make the assumption, tacitly made in former presentations, that these distances are equal, in which case both migrating ions meet half way between their original positions, a glance at Fig. 2 shows, that after electrolysis, that portion of liquid which borders on the anode will contain half an equivalent less cations than before. The converse is of course true for the other portion which is in contact with the cathode. By equivalent is understood the quantity of the component liberated.

If the two ions do not move through equal distances—that

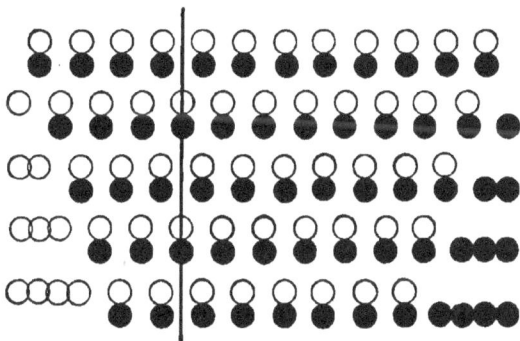

Fig. 2

is, if they do not meet each other half way—then the side of the liquid in which the more rapidly moving ion makes its appearance will be increased by more than half an equivalent of it, and diminished by less than half an equivalent of the other ion. Fig. 3 shows this for the case when the anion moves $\frac{1}{3}$, the cation $\frac{2}{3}$ of the distance. The anode side of the liquid contains $\frac{1}{3}$ of an equivalent more anions and $\frac{2}{3}$ of an equivalent less cations after the decomposition than before. The other side shows the converse relation.

This result evidently holds generally. If one ion moves through $1/n$ the distance, and the other $\frac{n-1}{n}$, then the side of the liquid in which the former appears will contain $1/n$ equivalent more of it and $\frac{n-1}{n}$ equivalent less of the other ion. The converse relation will hold for the other side of the electrolyte.

The first experiments to determine the transference of ions quantitively, were made by Faraday.[*] He took up the subject, however, only as a side issue, and confined himself to two elec-

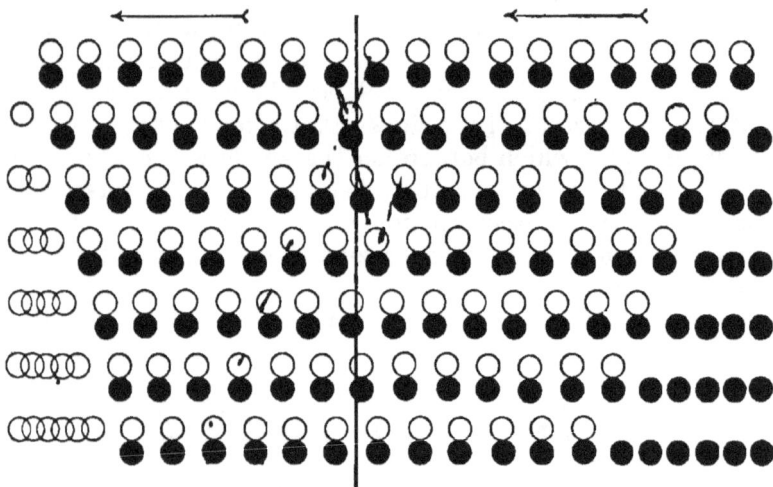

Fig. 3

trolytes, dilute sulphuric acid, and a solution of sodium sulphate. Two pairs of cups were filled respectively with definite amounts of these two liquids, and each pair connected together by means of asbestos. Both were then introduced into the same circuit, and after electrolysis had continued for some time, the asbestos was withdrawn, and the contents of the cups subjected to analysis. It is clear that this method is very defective, and that no accurate results are to be expected with it. The results which Faraday obtained in two series of experiments show this sufficiently. In the case of the sodium salt, he determined only the sulphuric acid set free, and tacitly assumed that half of it had been transferred.

---

[*] *Exp. Research.*, § 525–530 ; Pogg. *Ann.*, **32**, 436.

Messrs. Daniell and A. Miller,[*] in their beautiful investigations on the electrolysis of salts, were led to devote greater attention to the subject of transference. They effected the separation of the liquid by the introduction of a membrane. They filled the two cells into which the vessel was divided with accurately determined quantities of the aqueous solution of the electrolyte, and investigated each after the galvanic decomposition had taken place.

The results which they obtained are very striking. When, namely, copper or zinc sulphate was chosen as electrolyte, they found after electrolysis, exactly the same amount of metal in the cell containing the cathode as they had originally introduced. The quantity of reduced metal, increased by the quantity still dissolved in the liquid, amounted to exactly as much as was present before the electrolysis. According to this, copper and zinc do not migrate at all during electrolysis. Their anion $\ddot{S}$ traverses the whole distance. An ammonium salt (salammoniac) gave the same result; the complex cation NH_4 is to be classed with the two preceding. They found a transference of the cation with the salts potassium sulphate, barium nitrate, and magnesium sulphate. For potassium it amounted to $\frac{1}{5}$, for barium $\frac{1}{4}$, and for magnesium $\frac{1}{12}$ equivalent. The authors conclude from their experiments that those metals which decompose water at ordinary temperature, or whose oxides are easily soluble in water, are subject to a progressive transference in the voltaic cell from anode to cathode during electrolysis, while those which do not possess so strong an affinity for oxygen retain their place. They found a transference of all anions, even the weakest ones, such as WO_4 and CO_3.

In the translation of their article in Poggendorff's *Annalen*, the direct numerical results of the individual experiments are not given completely. The accuracy of the method cannot therefore be judged. It would appear, however, that it was not satisfactory, as the results are given only in round numbers. Furthermore, it is expressly stated that the experiments are not strictly comparable, and that the figures given cannot be regarded as absolute determinations of the transferred quantity of each metal in the cell.

The introduction of the membrane entails of necessity two

* Pogg. *Ann.*, **64**, 18.

evils. The lesser lies in the fact that the contents of each cell cannot be completely removed after the electrolysis, as some of the solution remains either in the diaphragm or comes through from the other cell. The more serious is a result of the inexplicable phenomenon, that the quantity of liquid in the negative cell increases, and in the positive cell diminishes, in these experiments. This was frequently noticed by Daniell, and has been very recently more carefully investigated by Wiedemann.* The latter regards it as a motion of the liquid mass as a whole from anode to cathode, and finds it very marked in copper and zinc vitriol solutions. It seems doubly striking, therefore, that Daniell and Miller found the quantity of copper unchanged in the negative cell, since an increase should have occurred as a result of this motion.

As proof that the diaphragm offers no obstruction to the progress of the ions, the authors cite the phenomenon familiar to electrotypers, that, in a copper vitriol solution, the liquid about the negative pole becomes weaker in copper and finally exhausted, when the negative pole is placed in the upper and the positive pole in the lower layers of the solution. They tried a similar experiment by filling a long tube, provided with two upright arms, with a strong solution of copper sulphate, and connected it by means of copper strips to a battery. The liquid in the negative arm became noticeably lighter colored, while, on the other hand, that in the positive arm became darker. From this they concluded that the oxysulphur ion ($\ddot{S}$), which separated out at the latter place, dissolved copper from the anode, but that this copper could not migrate to the cathode so as to replace the metal precipitated there.

This same phenomenon was reported at nearly the same time by numerous physicists, and introduced into discussions on the process of electrolysis. Pouillet† describes it in a gold solution which was contained in a U-shaped tube. After a current had passed a sufficiently long time, he found the solution in the negative arm almost completely deprived of its gold, while that in the positive arm still contained its original gold contents. He concludes from this "that in the decomposition of gold chloride, and therefore all metal salts, the positive pole has no

---

* Pogg. *Ann.*, **87**, 321.
† *Ibid.*, **65**, 474.

decomposing action, that all chemical force resides in the negative pole, that this takes up the gold and sends the chlorine by a series of successive decompositions and recombinations to the positive pole, to be there set free." "If both poles acted," he adds, "the metal separated at the negative pole would be of double origin; one-half would be directly precipitated, the other would come from the positive pole; both arms of the tube would then become weaker in gold to the same extent during the whole duration of the process."

Besides the physicists mentioned, Smee * also discusses the phenomenon.

It is astonishing how this simple experiment has been so generally misunderstood. The dilution which the solution undergoes at the negative pole proves in no way that the metal does not migrate during electrolysis. We can convince ourselves of this at once by glancing back at Fig. 2 or 3. The cation, in the above case, is a solid body in the free state, and, as such, leaves the solvent by the separation produced by the current. Fig. 2 is drawn on the assumption that the ions move through equal distances, and shows that the cathode side is increased after electrolysis by $\frac{1}{4}$ equivalent of cations. Now as one equivalent becomes solid, the solution is thereby diminished by $\frac{1}{4}$ equivalent—that is, diluted by $\frac{1}{4}$ equivalent of the salt. Dilution must therefore occur at the negative pole, even if the cation migrates; and it must evidently do so in all cases, as long as the cation does not migrate alone and the anion remain at rest. Only in this one case will the original concentration remain the same at the cathode.

This very dilution which the cation suffers at the negative pole where the cation leaves the solution, can be very advantageously used to determine the transference quantitatively. An exact separation of the electrolyte is easily effected without the introduction of asbestos or of a diaphragm.

Fig. 4 represents a simple apparatus which I have constructed for this purpose, and which was used in the experiments described below.

A glass cylinder, which contains the solution of the electrolyte, is composed of two parts—a larger, a, and a smaller, b. The former is cemented to a vessel, c, preferably of porcelain,

* Pogg. Ann., **65**, 473.

and contains the anode *d*. This has the form of a circular perforated plate, and is made of metal, the salt of which is to be electrolyzed. The support fastened to its centre passes through a small cork in a glass plate and permits connection with the galvanic cell. This plate forms the base of the cylinder, and is held in place by a cover which screws on. The anode is not permitted to lie on the bottom, but is placed a little higher up, so that the concentrated solution, which forms at its surface during electrolysis, can. flow down through the holes. The smaller part of the cylinder *b* is closed above .by a similar perforated glass plate, provided with a cork, and contains the cathode *e*, likewise fastened to a support which projects outward. The cathode must be given a different form from the anode. If it consists of a horizontal plate, the metal deposited by the current on the under surface cannot hold. It falls down and sets the liquid in motion. In order to prevent this, a metal cone is used as cathode, which is fixed with its apex at the centre of a horizontal circular glass plate. The glass plate is much smaller than the cross-section of the cylinder, and so chosen that points in its circumference are approximately equally distant from the base and from the apex of the cone. By this device all parts of the surface of the cone are nearly the same distance from the anode, and the deposited metal distributes itself nearly uniformly over the whole. The base of the cone presses closely against the plate forming the cover. Its height is so chosen that the glass plate *f* comes about in the middle of the cylinder. The cone and support are made preferably of platinum or gold. Failing these, silver may be used, which is what I was obliged to use.

When an experiment is to be made, the lower cemented part

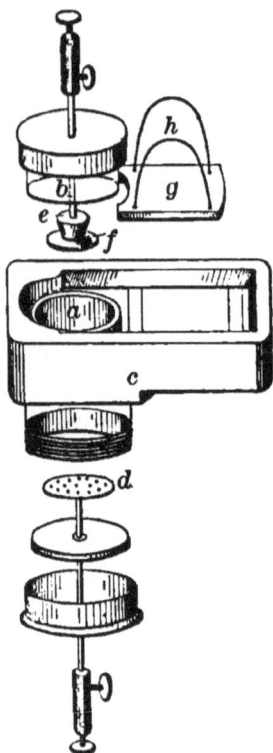

Fig. 4

of the cylinder, together with the vessel. c, is first filled with the solution. The same is done with the upper part in which the cathode is placed, care being taken that no air-bubbles remain in the inside. By means of a glass plate, g, which is ground on the open end of this cylinder, a definite quantity of liquid can be measured off. When this has been done the cylinder is inverted and, together with the glass plate, placed in the vessel c beside the cylinder a. For convenience of manipulation, a silver wire, h, passes through four holes in the corners of the plate, thereby forming two handles. The vessel c is just large enough to permit the cylinder a and the glass plate g to rest side by side on the bottom. The cylinder a, moreover, is so cemented in, that its upper edge projects above the bottom by just the thickness of the glass plate, so that it lies in the same plane with the upper surface of the latter. The smaller cylinder filled with solution can, therefore, be easily slid along from the plate on to the lower cylinder, thus forming a single cylinder. In this position its contents are supported by the atmospheric pressure.

The solution contained in the cylinder undergoes a change only at the electrodes during electrolysis. The liquid around the anode becomes more concentrated and, therefore, remains in the lower part ; the solution around the cathode becomes diluter and collects on the cover. When the current has decomposed a sufficient quantity, the upper cylinder is slid back again onto the glass plate, and taken out. The outside is cleaned from adhering liquid, and the contents carefully poured into another vessel for analysis. If the upper cylinder be now filled with the original solution and this quantity likewise analyzed, one has, with the quantity of metal deposited, all data necessary for computing the transference.

The cathode projects, intentionally, only to the centre of the upper cylinder in order that the liquid at the opening shall remain unchanged, and the mixing with the liquid in the vessel c, which occurs at this place on sliding the cylinder back on to the glass plate, shall occasion no error. To prevent the liquid in c from becoming concentrated by evaporation during electrolysis, the apparatus is set into a ground-glass plate, i, and covered with a bell-glass during the experiment. Fig. 5 represents in cross-section the apparatus completely set up. The dimensions of my apparatus are as follows: The inside diam-

eter of the cylinder measures 54 mm.; the height of the lower
part, 70 mm.; that of the upper part, 25 mm., both inside measurements. The glass
is $4\frac{1}{2}$ mm. in thickness, as it must be taken
rather thick. As the
cathode extends only
to the middle of the
upper cylinder, the ef-
fect of diffusion is de-
stroyed in our experi-
ments. During the
comparatively short
duration of the elec-
trolysis, this will be
active only between
the layers in the upper
cylinder, and will have
no effect on the mass
in the lower one; it
can, therefore, be the

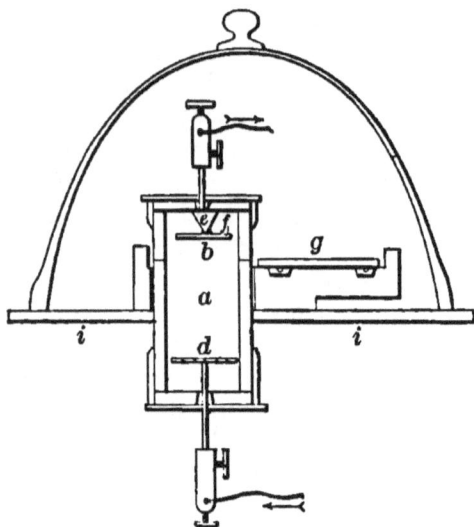

Fig. 5

cause of no error. Moreover, the motion, which, according to
Wiedemann, the electrolyte as a whole experiences from the
anode to the cathode, cannot vitiate our results, as it cannot
take place under the above conditions. The only error, so far
as I can see, which enters into my method and cannot be
avoided, arises from the fact that the metal which is separated
out by the current has a different volume from that of the salt
which is carried away from the upper part. This change in
volume is replaced by liquid flowing in or out. The values
which we obtain for the transference will be incorrect by the
contents of this quantity of liquid. Our error is, however, very
insignificant, and may be at least approximately computed.
We shall see that even in the case of very concentrated solutions it does not amount to as much as the unavoidable error
of analysis. This will be all the more true in the case of the
dilute solutions, for, as is readily seen, the error must, in general, diminish proportional to the dilution.

Besides the apparatus, a voltameter was introduced into the
circuit. I chose for this purpose the convenient and accurate
arrangement described by Poggendorff, called a silver voltam-

eter. A silver dish, which served as cathode, contained a solution of silver nitrate, into which dipped a silver plate as anode. The latter was wrapped around with a linen cover to prevent little particles, which easily come off during solution of the anode by the liberated anion N, from falling into the dish, and thus increasing the weight of the reduced silver. The first salt which I decomposed was copper sulphate, with which Daniell and Miller also worked, and which possesses special interest on account of its application in galvanoplasty. It is the most convenient electrolyte for our experiments, for, as is well known, copper deposits coherently, and consequently adheres firmly to the surface of the silver cone.

I. COPPER SULPHATE

The solution which was subjected to electrolysis was prepared by diluting a more concentrated one to about twice its volume. Its specific gravity at $4.9°$ C. was 1.1036, and it contained 1 part $\dot{S}\ \ddot{C}u$ to 9.56 parts water, or 1 part $(\dot{S}\ \ddot{C}u+5\ \dot{H})$* to 5.75 parts water.

Experiment A

The electrolysis was carried out at the temperature $4.7°$ C., and was effected by means of a small Grove cell. The current continued four hours and reduced 1.008 gr. Ag in the voltameter, or 0.0042 gr. Ag per minute.

This quantity of silver is equivalent to 0.2955 gr. Cu.

There was deposited on the silver cone, however, 0.2975 gr. Cu.

The difference, 0.002 gr., arises without doubt from an oxydation of the copper ; we base all calculations on values deduced from results obtained by the silver voltameter.

The solution about the cathode contained :

Before electrolysis......... 2.8543 gr. $\ddot{C}u$
After   "   .... 2.5897 "   "

It was therefore diluted by an amount 0.2646 gr. $\ddot{C}u=0.2112\ Cu$.

The $\ddot{C}u$ was precipitated in the usual way by caustic potash, from a boiling solution.

The amount of transferred copper is therefore

---

* [*Nomenclature introduced by Berzelius. The dots over an element represent the number of attached oxygen equivalents. Gmelin's equivalents ; i.e.,* $H=1$, $O=8$, $S=16$, *etc., are used throughout.*]

0.2955
—0.2112
———————
0.0843 gr., *i.e.*, $\dfrac{843}{2955} = 28.5$ per cent. equivalent.

Our experiment gives a totally different result from that obtained by Messrs. Daniell and Miller. According to their results, the solution in the upper cylinder should have lost 0.2955 gr. Cu during the electrolysis.

We will next consider whether the transference remains constant for all current strengths. To obtain an answer to this question, the above solution was subjected to the action of a weak and of a strong current.

Experiment B

The current from a Grove cell was so cut down by the introduction of a long thin German-silver wire, that at a temperature of 5.3° C. it reduced 1.2273 gr. Ag in 18 hours and 4 minutes, or 0.00113 gr. Ag per minute.

The quantity of silver corresponds to 0.3597 gr. Cu.
There was deposited on the silver cone 0.3587 gr. Cu.
The solution about the cathode contained :

Before electrolysis....... 2.8543 gr. Cu
After   "   .... 2.535   "   "

It was therefore diluted...... 0.3193 gr. Cu, or 0.2549 gr. Cu.

The quantity of transferred copper is therefore

0.3597
—0.2549
———————
0.1048 gr., or $\dfrac{1048}{3597} = 29.1$ per cent. equivalent.

Experiment C

The current from three Grove cells reduced at 6.5° C. 1.1503 gr. Ag in 2 hours, or 0.00958 gr. Ag a minute.
This quantity of silver corresponds to 0.3372 gr. Cu.
There was deposited on the silver cone 0.3374 gr. Cu.

The solution around the cathode contained :

Before electrolysis....... 2.8543 gr. Ċu
After   "   .... 2.5541 "   "

It lost therefore............ 0.3002 gr. Ċu, or 0.2396 gr. Cu.

The quantity of transferred copper is therefore

0.3372
−0.2396
―――――
0.0976 gr., or $\dfrac{976}{3372} = 28.9$ per cent. equivalent.

If we tabulate the results of these experiments :

CURRENT	TRANSFERENCE
113	29.1
420	28.5
958	28.9 per cent.
Mean..	28.8 per cent.

there can be no doubt that the transference is independent of the intensity of the current. I have always avoided using very large currents, as the rise in temperature which they produce in the solution is disturbing. The immediate effect of this on our data is easily obviated by not removing the electrolyzed solution for analysis immediately after breaking the current, but allowing it to first return to the temperature of the surroundings. On the other hand, an indirect disturbance of the rise of temperature cannot be so easily overcome. This consists in the evolution of a quantity of little air-bubbles which usually cover the surface of the glass plate under the cathode, and which cannot be removed. That these little bubbles are not hydrogen gas is clear from the place where they appear. If large currents are to be used, it is judicious to free the solution as far as possible from absorbed air, before filling the apparatus ; this is most easily done under an air-pump.

The second question which we must consider has reference to the influence of the concentration on the transference. Six solutions of copper sulphate of very different concentration were subjected to electrolysis.

Solution I

A concentrated solution was diluted just sufficiently so that a separation of salt on the anode was not to be feared. It had at 4.5° C. a specific gravity of 1.1521, and contained 1 part $\overset{..}{S}\overset{.}{C}u$ to 6.35 parts water, or 1 part $(\overset{..}{S}\overset{.}{C}u + 5\ \overset{..}{H})$ to 3.69 parts water. The current from a Grove cell deposited, at 5.5° C., 1.0783 gr. Ag in 4 hours. This corresponds to 0.3161 gr. Cu. On the silver cone there was 0.3168 gr. Cu. The solution around the cathode contained :

Before electrolysis.......	4.2591 gr. $\overset{.}{C}u$	
After "	3.9725 "	"
It lost.....................	0.2866 gr. $\overset{.}{C}u$, or 0.2288 gr. Cu.	

The amount of transferred copper is therefore

$$0.3161$$
$$-0.2288$$

$$0.0873 \text{ gr., or } \frac{873}{3161} = 27.6 \text{ per cent.}$$

The solution first electrolyzed, which gave 28.8 per cent. transference, served as Solution II.

Solution III

Specific gravity at 3.6° C.: 1.0553.

It contained 1 part $\overset{..}{S}\overset{.}{C}u$ to 18.08 parts water, or 1 part $(\overset{..}{S}\overset{.}{C}u + 5\ \overset{..}{H})$ to 11.19 parts water.

The current from one Grove cell deposited 0.8601 gr. Ag in 5 hours, 45 minutes, at 5.5° C. This corresponds to 0.2521 gr. Cu.

There was 0.2520 gr. Cu on the silver cone.

The solution around the cathode contained :

Before electrolysis.....	1.5026 gr. $\overset{.}{C}u$	
After "	1.2895 "	"
It lost...................	0.2131 gr. $\overset{.}{C}u$, or 0.1701 gr. Cu.	

The amount of transferred copper is therefore

$$0.2521$$
$$-0.1701$$

$$0.0820 \text{ gr., or } \frac{820}{2521} = 32.5 \text{ per cent.}$$

Solution IV

Specific gravity at 3° C.: 1.0254.

It contained 1 part $\ddot{S}\dot{C}u$ to 39.67 parts water, or 1 part $(\ddot{S}Cu+5\ H)$ to 24.99 parts water.

The current from two Grove cells deposited at 4.5° C., 0.6969 gr. Ag in 5 hours : this is equivalent to 0.2034 gr. Cu.

The copper which covered the silver cone could no longer be weighed, as in this dilute solution the larger part of it was spongy.

The solution around the cathode contained :

Before electrolysis..... 0.6765 gr. Ċu
After   "   ..... 0.5118 "   "

It lost................... 0.1647 gr. Cu, or 0.1315 gr. Cu.

Hence the transference of the copper is
$$0.2043$$
$$-0.1315$$
$$\overline{0.0728}\ \text{gr., or}\ \frac{728}{2043}=35.6\ \text{per cent.}$$

Solution V

Specific gravity at 4.8° C.: 1.0135.

It contained 1 part SCu to 76.88 parts water, or 1 part $(\ddot{S}Cu+5\ \ddot{H})$ to 48.75 parts water.

The current of one Grove cell reduced 0.3592 gr. Ag at 4.3° C. This corresponds to 0.1053 gr. Cu.

The copper on the silver cone was spongy. The solution about the cathode contained :

Before electrolysis..... 0.3617 gr. Ċu.
After   "   ..... 0.2758 "   "

It lost................... 0.0859 gr. Ċu, or 0.0686 gr. Cu.

Hence the transference of copper is
$$0.1053$$
$$-0.0686$$
$$\overline{0.0367}\ \text{gr., or}\ \frac{367}{1053}=34.9\ \text{per cent.}$$

Solution VI

Specific gravity at 4.4° C.: 1.0071.

It contained 1 part S̈Ċu to 148.3 parts water, or 1 part (S̈Cu + 5 Ḧ) to 94.5 parts water.

The current from one Grove cell reduced 0.3850 gr. Ag in 16 hours, 25 minutes, at 4.4° C. This corresponds to 0.1131 gr. Cu.

The copper on the silver cone was spongy.

The solution around the cathode contained :

Before electrolysis..... 0.1867 gr. Ċu
After   "   ..... 0.0964 "   "

It lost.................. 0.0903 gr. Ċu, or 0.0721 gr. Cu.

The transference of copper is

$$\begin{array}{r} 0.1131 \\ -0.0721 \\ \hline 0.0410 \end{array}$$ gr., or $\dfrac{410}{1131} = 36.2$ per cent.

Let us tabulate the separate results together for inspection.

NO.	SP GR.	CONTENTS OF SOLUTION			TRANSFERENCE OF COPPER	
I	1.1521	1 pt. S̈ Ċu to	6.35 pts.	H	27.6	per cent.
II	1.1036	" " "	9.56 "	"	28.8	" "
III	1.0553	" " "	18.08 "	"	32.5	" "
IV	1.0254	" " "	39.67 "	"	35.5	mean, 35.6 per cent.
V	1.0135	" " " ·	76.88 "	"	34.9	
VI	1.0071	" " "	148.3 "	"	36.2	

The transference numbers still require the small correction which I pointed out above. We can only estimate this approximately, as we cannot determine with our method, throughout how large a portion of the solution the dilution extends. The dilute solution, which can be easily followed with the eye during the electrolysis, forms directly on the surface of the silver cone, glides upward along it, and collects under the cover. To obtain at least an idea of the amount of this correction, I will calculate it for Solution I, under a definite assumption which will not be far from the truth.

The liquid at the cathode lost 0.2866 gr. $\dot{C}u$, or 0.5762 gr. $\ddot{S}Cu$. Suppose this loss extends over such a mass, x, of the liquid that a solution of concentration II thereby results. Before electrolysis the quantity x contains

$$\frac{6.35}{7.35} x \text{ water, and } \frac{1}{7.35} x \ \ddot{S}\dot{C}u.$$

After electrolysis it will contain

$$\left(\frac{1}{7.35} x - 0.5762\right) \text{gr.} \ \ddot{S} \ \dot{C}u,$$

and be of concentration II ; it will therefore contain.

$$\frac{6.35}{7.35 \times 9.56} x \ \ddot{S} \ \dot{C}u \text{ to } \frac{6.35}{7.35} x \text{ water.}$$

The mass sought is therefore obtained from the equation

$$\frac{6.35}{7.35 \times 9.56} x = \frac{1}{7.35} x - 0.5762,$$

and equals $x = 12.616$ gr. Before electrolysis this mass has the volume

$$\frac{12.616}{1.1521} = 10.9504 \text{ c. cm.}$$

It loses 0.5762 gr. $\ddot{S} \ \dot{C}u$ by electrolysis, and the volume becomes $\frac{12.0398}{1.1036} = 10.9095$ c. cm. Hence the withdrawal of 0.5762 gr. $\ddot{S} \ \dot{C}u$ causes a diminution of volume of 0.0409 c. cm. According to *Marchand* and *Scherer*,* galvanically deposited copper has a density of 8.914. Hence the reduced 0.3161 gr. Cu occupies a volume of 0.0355 c. cm. The diminution exceeds the increase in volume by $0.0409 - 0.0355 = 0.0054$ c. cm. This volume is replaced by the solution flowing in. The latter weighs 0.0054×1.1521 gr. $= 0.0062$ gr., and contains 0.00042 gr. $\dot{C}u$.

Hence, even in case of this most concentrated solution, the error is of no account. This will be even more true in the other cases.

(The effect of the water on the amount of transference is evident from the experimental results. In proportion as the dilution increases, the transference of the cation $\dot{C}u$ increases and of the anion ($\ddot{S}$) decreases. \ In Solution IV the limit of this influence seems to be reached. From there on the numbers become nearly constant.

There still remains a third condition which can affect the

* *Gmelin*, iii., 374.

transference ; I mean temperature. Our apparatus allows us to work only at temperatures which we can give to the surrounding air.

A solution was prepared which had about the same concentration as Solution II.

Experiment D

During the electrolysis of this solution the temperature of the air varied from 21° to 18° C. The current from one Grove cell reduced 1,4247 gr. Ag in 4 hours 3 minutes. This corresponds to 0.4176 gr. Cu.

0.419 gr. Cu was found on the silver cone.

The solution about the cathode contained :

Before electrolysis.....	2.8921 gr. Ċu	
After "	2.5191 " "	
It lost...................	0.3730 gr. Ċu, or	0.2977 gr. Cu.

Hence the transference of the copper is

$$\frac{\begin{array}{r} 0.4176 \\ -0.2977 \end{array}}{0.1199 \text{ gr.}}, \text{ or } \frac{1199}{4176} = 28.7 \text{ per cent.}$$

The temperature has no effect between 4° and 21° C.*

Copper vitriol is a salt which crystallizes from aqueous solutions with five molecules of water. The remarkable influence which the amount of water exerts on the transference made the investigation of an anhydrous salt especially desirable. I chose

II. SILVER NITRATE

The salt was melted before dissolving in order to obtain it absolutely neutral. The solution did not react with litmus. It is not as convenient for electrolysis as copper sulphate, as the silver adheres firmly to the cone only when deposited from quite concentrated solutions, and with weak currents. Usually the dendritic crystals grow rapidly over the glass plate underneath the cathode and fall off.

* [This conclusion has not been verified by more recent experiments. See Loeb and Nernst, Zeit. für Phys. Chem., 2, 948, 1888; Bein, Wied. Ann., 46, 29, 1892.]

I chose such currents that a sufficient amount of silver was reduced before it began to drop off. When this threatened to occur the electrolysis was stopped.

Solution I

Specific gravity at 11.1° C.: 1.3079.
    It contained 1 part N̈Ag to 2.48 parts water.
    The current reduced 1.2591 gr. Ag in 1¼ hours at a temperature of 11.2° C.
    The solution about the cathode gave :

        Before electrolysis... 17.4624 gr. ClAg
        After   "   ... 16.6796 "   "
                                 _____
    It lost................. 0.7828 gr. ClAg, or 0.5893 gr. Ag.

Hence the amount of the transferred silver is
                1.2591
               −0.5893
               _____
                0.6698 gr., or $\frac{6698}{12591} = 53.2$ per cent.

Solution II

Specific gravity at 19.2° C.: 1.2788.
    It contains 1 part N̈Ag to 2.735 parts water.
    The current from one cell reduced 1.909 gr. Ag at 19° C.
    The solution at the cathode gave :

        Before electrolysis... 15.9364 gr. ClAg
        After   "   ... 14.7233 "   "
                                 _____
    The loss is .... 1.2131 gr. ClAg, or 0.9132 gr. Ag.

The transference of silver is therefore
                1.909
               −0.9132
               _____
                0.9958 gr., or $\frac{9958}{19090} = 52.2$ per cent.

Solution III

Specific gravity at 18.4° C.: 1.1534.
    It contains 1 part N̈Ag to 5.18 parts water.

The current from one cell reduced 1.1124 gr. Ag in 1 hour 21½ minutes at a temperature of 18.4° C.
The solution about the cathode gave :

Before electrolysis... 8.6883 gr. ClAg
After   "   ... 7.9569 "   "

The loss is.............. 0.7314 gr. ClAg, or 0.5506 gr. Ag.

Hence the amount of transferred silver is

1.1124
−0.5506

0.5618 gr., or $\dfrac{5618}{11124}=50.5$ per cent.

Solution IV

Specific gravity at 18.8° C.: 1.0774.
It contained 1 part N̈Ag to 10.38 parts water.
The current from two cells reduced 0.4541 gr. Ag in half an hour at 18.8° C.
The solution about the cathode gave :

Before electrolysis... 4.4156 gr. ClAg
After   "   ... 4.1080 "   "

The loss is.............. 0.3076 gr. ClAg, or 0.2316 gr. Ag.

Hence the amount of transferred silver is

0.4541
−0.2316

0.2225 gr., or $\dfrac{2225}{4541}=49$ per cent.

Solution V

Specific gravity at 19.2° C.: 1.0558.
It contained 1 part N̈Ag to 14.5 parts water.
The current from two cells reduced 0.3937 gr. Ag in 25 minutes at 19.2° C.
The solution about the cathode gave :

Before electrolysis... 3.1731 gr. ClAg
After   "   ... 2.8985 "   "

The loss is.............. 0.2746 gr. ClAg, or 0.2067 gr. Ag.

The amount of transferred silver is therefore

$$\begin{array}{r} 0.3937 \\ -0.2067 \\ \hline 0.1870 \end{array}$$ gr., or $\dfrac{1870}{3937} = 47.5$ per cent.

Solution VI

Specific gravity at 18.4° C.: 1.0343.
It contains 1 part N̈Ag to 23.63 parts water.
The current from two elements reduced 0.3208 gr. Ag in half an hour at 18.4° C.
The solution at the cathode gave:

Before electrolysis...	1.9605 gr. ClAg	
After " ...	1.7358 " "	
The loss is..............	0.2247 gr. ClAg, or 0.1691 gr. Ag.

Hence the amount of transferred silver is

$$\begin{array}{r} 0.3028 \\ -0.1691 \\ \hline 0.1517 \end{array}$$ gr., or $\dfrac{1517}{3208} = 47.3$ per cent.

Solution VII

Specific gravity at 18.5° C.: 1.0166.
It contains 1 part N̈Ag to 49.44 parts water.
The current from two cells reduced 0.2470 gr. Ag in 45½ minutes at 18.5° C.
The solution around the cathode gave:

Before electrolysis...	0.9485 gr. ClAg	
After " ...	0.7758 " "	
The loss is..............	0.1727 gr. ClAg, or 0.1300 gr. Ag.

The amount of transferred silver is therefore

$$\begin{array}{r} 0.2470 \\ -0.1300 \\ \hline 0.1170 \end{array}$$ gr., or $\dfrac{1170}{2470} = 47.4$ per cent.

Solution VIII

Specific gravity at 18.6° C.: 1.0076.

It contains 1 part ÑAg to 104.6 parts water.

The current from three elements reduced 0.1888 gr. Ag in 41 minutes at 18.6° C.

In this very dilute solution the silver separated out on the silver cone at first black and spongy, as described by Poggendorff,* and became afterwards yellowish-white and crystalline.

The solution about the cathode gave :

Before electrolysis... 0.4515 gr. ClAg
After   "   ... 0.3197 "   "
The loss is............. 0.1318 gr. ClAg, or 0.0992 gr. Ag.

The amount of transferred silver is
$$0.1888$$
$$-0.0992$$
$$0.0896 \text{ gr., or } \frac{896}{1888} = 47.4 \text{ per cent.}$$

Solution IX

Specific gravity at 9.6° C.: 1.0044.

It contains 1 part ÑAg to 247.3 parts water.

The current from four elements reduced 0.0863 gr. Ag in 1 hour 3 minutes at 9.6° C.

The solution about the cathode gave :

Before electrolysis... 0.1916 gr. ClAg
After   "   ... 0.1316 " · "
The loss is............. 0.0600 gr. ClAg, or 0.0452 gr. Ag.

Hence the transference of the silver is
$$0.0863$$
$$-0.0452$$
$$0.0411 \text{ gr., or } \frac{411}{863} = 47.6 \text{ per cent.}$$

We will again tabulate the results obtained with the nine different solutions.

---

* Pogg. *Ann.*, **75**, 338.

NO.	SP. GR.	CONTENTS		TRANSFERENCE OF SILVER
I	1.3079	1 pt. N̈Ag. to	2.48 pts. Ḧ	53.2 per cent.
II	1.2788	" " "	2.73 "	52.2 " "
III	1.1534	" - " "	5.18 "	50.5 " "
IV	1.0774	" " "	10.38 "	49. " "
V	1.0558	" " "	14.5 "	47.5
VI	1.0343	" " "	23.63 "	47.3
VII	1.0166	" " "	49.44 "	47.4 47.44 mean
VIII	1.0076	" " "	104.6 "	47.4 per cent.
IX	1.0044	" " "	247.3 "	47.6

The correction which ought to be applied to these figures is here again, even for Solution I, so small that it falls within the experimental error. If we make the same assumption as in the case of copper vitriol, it amounts to 0.0005 gr. for the 0.6698 gr. of transferred silver. The effect of water in the case of silver nitrate is opposite to that in the case of copper vitriol. The transference of the cation Ag diminishes, that of the anion N̈ increases, with increasing amount of the solvent. The effect of the water reaches a limit in Solution V. Greater dilution does not further change the value.

In the two above salts, the ions are all different substances. I now investigated compounds of the same cation with different anions, and chose for this purpose silver sulphate and silver acetate. Both of these salts are difficultly soluble in water, but still sufficiently so to give accurate results for our purpose.

III. SILVER SULPHATE

Experiment A

Specific gravity of the solution at 15° C.: 1.0078.

The solution contained 1 part S̈Ag to 123 parts water.

The current from four elements reduced 0.1099 gr. Ag in 24 minutes at 15° C.

The solution about the cathode gave :

Before electrolysis... 0.4166 gr. ClAg

After   "   ... 0.3358 "   "

The loss is.............. 0.0808 gr. ClAg, or 0.0608 gr. Ag.

The quantity of transferred silver is therefore
$$
\begin{array}{r}
0.1099 \\
-0.0608 \\
\hline
0.0491
\end{array}
$$
gr., or $\dfrac{491}{1099} = 44.67$ per cent.

Experiment B

The current from four elements reduced 0.1127 gr. Ag in 25 minutes.

The solution around the cathode gave:

Before electrolysis... 0.4090 gr. ClAg
After   "   ... 0.3261 "   "
The loss is.............. 0.0829 gr. ClAg, or 0.624 gr. Ag.

Hence the amount of transferred silver is
$$
\begin{array}{r}
0.1127 \\
-0.0624 \\
\hline
0.0503
\end{array}
$$
gr., or $\dfrac{503}{1127} = 44.63$ per cent.

Experiment C

The current from four elements reduced 0.1108 gr. Ag in 23½ minutes at 19.4° C.

The solution around the cathode gave:

Before electrolysis... 0.3539 gr. ClAg
After   "   ... 0.2720 "   "
The loss is.............. 0.0819 gr. ClAg, or 0.0616 gr. Ag.

The transference of silver is therefore
$$
\begin{array}{r}
0.1108 \\
-0.0616 \\
\hline
0.0492
\end{array}
$$
gr., or $\dfrac{492}{1108} = 44.4$ per cent.

The results of the three experiments:

         44.67 per cent.
         44.63 "   "
         44.4 "   "

give the mean........ 44.57 per cent.

IV. SILVER ACETATE

Experiment A

Specific gravity of the solution at 14° C.: 1.0060.

It contained 1 part Äc Ag * to 126.7 parts water.

The current from four elements reduced 0.2197 gr. Ag in 1 hour 21 minutes at 14° C.

The solution at the cathode gave :

Before electrolysis... 0.3736 gr. ClAg
After   "   ... 0.2631 "   "

The loss is.............. 0.1105 gr. ClAg, or 0.0832 gr. Ag.

Hence the transference of silver is

$$\begin{array}{r} 0.2197 \\ -0.0832 \\ \hline 0.1365 \end{array} \text{ gr., or } \frac{1365}{2197} = 62.13 \text{ per cent.}$$

Experiment B

The current from four elements reduced 0.1892 gr. Ag in 1 hour 7 minutes at 15° C.

The solution at the cathode gave : ⁻

Before electrolysis... 0.3656 gr. ClAg
After   "   ... 0.2728 "   "

The loss is.............. 0.0928 gr. ClAg, or 0.0699 gr. Ag.

The amount of transferred silver is

$$\begin{array}{r} 0.1892 \\ -0.0699 \\ \hline 0.1193 \end{array} \text{ gr., or } \frac{1193}{1893} = 63 \text{ per cent.}$$

Experiment C

Specific gravity at 15° C.: 1.0045.

The current from four elements reduced 0.1718 gr. Ag in 1 hour 13 minutes at 15° C.

The solution at the cathode gave :

Before electrolysis...... 0.2825 gr. ClAg
After   "   ..... 0.1977 "   "

The loss is............... 0.0848 gr. ClAg, or 0.0638 gr. Ag.

* [Äc is equivalent to $C_2 H_4 O_2$ minus $\frac{1}{2}H_2O$ in modern notation.]

The amount of transferred silver is

$$\begin{array}{r} 0.1718 \\ -0.0638 \\ \hline 0.1080 \end{array}$$

gr., or $\dfrac{1080}{1718} = 62.86$ per cent.

From the results of these three experiments,

$$\begin{array}{r} 62.13 \text{ per cent.} \\ 63 \quad `` \quad `` \\ 62.86 \quad `` \quad `` \\ \hline \end{array}$$

we obtain the mean... 62.66 per cent.

If we glance at the values obtained with the three silver salts, it is at once evident that the same cation migrates by different amounts when in combination with different anions, the condition of the solutions remaining otherwise the same.

With Ag(Äc) the transference of $\begin{array}{l} \text{Ag is } 62.6 \text{ per cent.} \\ \text{Äc } `` 37.4 \quad `` \quad `` \end{array}$

" Ag(N̈) " " " $\begin{array}{l} \text{Ag } `` 47.4 \quad `` \quad `` \\ \text{N̈ } `` 52.6 \quad `` \quad `` \end{array}$

" Ag(S̈) " " " $\begin{array}{l} \text{Ag } `` 44.6 \quad `` \quad `` \\ \text{S̈ } `` 55.4 \quad `` \quad `` \end{array}$

If the explanation of the transference numbers which we gave at the beginning of this paper is correct, then the distances traversed during electrolysis by Ag and Äc, Ag and N̈, and Ag and S̈, are in the ratio respectively of,

$$\begin{array}{r} 100 : 59.7 \\ 100 : 110.9 \\ 100 : 124.2 \end{array}$$

In these numbers a relation to chemical affinity is unmistakable. Of the three anions with which we are concerned, every chemist regards the Äc as the weakest, the S̈ as the strongest.

The same relation is evident if we compare the transference numbers of (S̈) Cu and (S̈) Ag. In the first of these two electrolytes which contain the same anion, the migration of the S̈ is 64.4 per cent. and of the Cu 35.6 per cent., while at the same concentration, the migration of S̈ in the second electrolyte is 55.4 per cent. and of Ag 44.6 per cent. The relative distances traversed are therefore:

For S̈ and Cu : 100 and 55.3
For S̈ and Ag : 100 and 80.5

In order to explain the relation indicated, the following consideration naturally offers itself. Of several anions in combination with the same cation, we will consider that one the most electro-negative which moves the greatest distance towards the anode. The analogous relation holds for several cations present with the same anion. The farther apart two substances stand from each other in the voltaic series, the stronger appears their chemical affinity. We might therefore look for a measure of chemical affinity in the distances through which the anions migrate during electrolysis. At present, however, I am far from ready to assign this significance to the above figures. When we consider that copper appears more positive than silver in its electrical aspect, and that the quantity of water exerts such a decided influence on the transference, a theory is by no means yet to be thought of.

I do not yet attempt to give an explanation of the influence of the water. Whatever hypothesis we propose for this, we must remember that the neutrality of the solution is not disturbed by the electrolysis—that free acid never makes its appearance at the cathode. We can determine the transference equally well in our experiments if we determine quantitively the acid in the solution about the cathode, or if we determine the base. I always prefer the former way in these investigations, when analytical methods permit the acid being more sharply determined.

In my experiments with the four salts, hydrogen was never separated out at the cathode along with the metal, although very dilute solutions have been electrolyzed. I took, of course, great care to make up neutral solutions and to exclude all free acid. Although Smee * obtained a different result in the electrolysis of copper sulphate, yet this is only apparently the case. Smee cites in support of the older view of galvanic decomposition, according to which water only is decomposed, and the metal is a result of the reduction caused by the liberated hydrogen, an experiment in which he decomposed a copper vitriol solution in a tall glass vessel with copper electrodes, the upper of which was negative and the lower positive. He observed copper to separate out on the former, at first in a compact, later in a spongy form, and then hydrogen evolved, while the

* Pogg. *Ann.*, **65**, 473.

upper portion of the solution gradually became completely colorless, and the lower positive electrode became covered with a thick layer of copper oxide. With the exception of the remark concerning the anode, I have always observed the same results when the cathode in my apparatus had the form of a horizontal plate. If we place it just at the surface, so that only its underside is in contact with the liquid, the copper appears immediately in the spongy form if the current is not too weak;→it soon falls off and leaves a surface of pure water in contact with the cathode, whence, of course, hydrogen must appear. This follows so clearly from Figs. 2 and 3 that a further discussion is superfluous. To avoid this result, my cathode was given the form of a cone.

Daniell* has already unquestionably proved the hydrogen which is evolved during the galvanic decomposition of aqueous solutions of the alkali or alkali earth salts, to be secondary. It is known that when salts of iron, manganese, cobalt, and nickel, even in perfectly neutral aqueous solution, conduct the current, hydrogen is set free simultaneously with the metals. Is this hydrogen likewise secondary? Nothing is easier than to answer this question. A solution of S̈ Fe, which was purified from free acid by repeated crystallization, was introduced into a circuit with a silver voltameter. An iron plate dipped in the solution as anode, and a platinum plate as cathode. The liquid about the latter is as neutral after the electrolysis as before. If the hydrogen be of secondary origin, it is evolved by a portion of the liberated iron decomposing the water by uniting with its oxygen. Hence ferrous oxide must be mixed with the reduced iron, and the total amount of F̈e corresponding will contain as much iron as is equivalent to the silver.

The two following experiments show this clearly:

Experiment A

The current from three elements reduced 3.672 gr. Ag in the silver voltameter, which is equivalent to 0.9537 gr. Fe. The deposited iron was dissolved in aqua regia and precipitated as F̈e by ammonia.

The F̈e weighed 1.3625 gr.; it contained, therefore, 0.9542 gr. Fe.

* Pogg. *Ann.*, *Ergänzbd*, i., 565.

LAWS OF ELECTROLYTIC CONDUCTION

Experiment B

The reduced silver weighed 3.0649 gr., and is equivalent to 0.7960 gr. Fe.
The Fe weighed 1.1375 gr., and contained 0.7966 gr. Fe.

We shall obtain further information of the effect of water on the migration if we substitute another solvent. Unfortunately, our choice in this direction is very limited. Absolute alcohol is the only liquid which can replace water, and this only in a few cases, as it dissolves only a few electrolytes.

Of our four salts, silver nitrate alone is soluble in absolute alcohol. At higher temperatures it is easily soluble; at lower temperatures, at which alone electrolysis can be carried out on account of the volatility of the alcohol, it is difficultly soluble. A solution saturated at a higher temperature contained at 5° C. only 1 part NAg in 30.86 parts of alcohol.

The solution which was electrolyzed was somewhat diluter. The glass plate fastened under the cathode by sealing-wax was replaced by an ivory plate which was screwed on. and the cylinder *a* was sealed into the vessel *c* with plaster of Paris. The solution conducted poorly.

Experiment A

The current from six elements reduced 0.2521 gr. Ag in 3 hours 32 minutes at 3.8° C.
The solution about the cathode gave :

Before electrolysis... 0.9181 gr. ClAg
After   "   ... 0.7264 "   "
The loss is............. 0.1917 gr. ClAg, or 0.1443 gr. Ag.

Hence the transference of silver is
0.2521
−0.1443
0.1078 gr., or $\frac{1078}{2521}=42.8$ per cent.

Experiment B

The current from six elements reduced 0.1367 gr. Ag in 2 hours 22 minutes at 5° C.

The solution about the cathode gave :

Before electrolysis... 0.8743 gr. ClAg
After   "   ... 0.7700 "   "
The loss is............. 0.1043 gr. ClAg, or 0.0785 gr. Ag.

The transference of silver is therefore

$$0.1367$$
$$-0.0785$$
$$0.0582 \text{ gr., or } \frac{582}{1367}=42.6 \text{ per cent.}$$

Hence in alcoholic solution the transference of Ag is 42.7 per cent. ; of $\overset{...}{N}$, 57.3 per cent. ; and the relative distances traversed are 100 and 134.2 respectively.

This result, which was not anticipated, indicates the great caution to be observed in the interpretation of our results. I intend next to study such salts as are easily soluble in absolute alcohol at low temperatures, and hope in the next communication to be able to present results on the salts of zinc, cadmium, iron, manganese, etc. With several of these hydrogen separates out at the cathode during the electrolysis. As the solution becomes diluted there, my apparatus can easily be adapted to this investigation by a slight modification. I then also intend to return to Daniell and Miller's method and to their discordant results.

BIOGRAPHICAL SKETCH

JOHANN WILHELM HITTORF was born in Bonn, May 27, 1824. He was made a member of the Philosophical Faculty of the Royal Academy of Münster in 1852, with the title of Professor of Chemistry and Physics, having previously occupied the position of Docent in the same institution. By the reorganization of the faculty in 1876, he was relieved of the instruction in chemistry. As professor of physics he continued work of instruction until 1890, when sickness compelled him to give up active work. He was then made Professor Emeritus, which position he still holds.

The valuable contributions which Hittorf has made to science, have made him an honored member of many societies. He is corresponding member of the Königliche Gesellschaft of Göttingen, Berlin, and Munich; foreign member of the Danish Academy of Copenhagen, and honorary member of the Manchester Literary and Philosophical Society, and of the London Physical Society. The degree of M.D. was conferred on him by the medical faculty of the University of Leipzig, and in 1897 he was honored with the Prussian order *pour le mérite* for science and arts. In 1898 he was elected Honorary President of the German Electrochemical Society.

Of Hittorf's published papers, most of which have appeared in Poggendorff's and Wiedemann's *Annalen* since 1847, the extended series of investigations on electrolysis, of which the above is the first, should first be mentioned. In 1864 the "Multiple Spectra" of the elements was established in an investigation with Plücker. In the years 1869–1874 a series of important papers appeared on the phenomena accompanying the passage of electricity through rarefied gases, and on the remarkable behavior of cathode rays.

In Chemistry may be mentioned an investigation on the allotropic forms of selenium and phosphorus, in which a new black metallic crystalline modification of the latter was discovered. Quite recently Hittorf has contributed several articles to the *Zeitschrift für Physikalische Chemie.*

ON THE CONDUCTIVITY OF ELECTRO-LYTES DISSOLVED IN WATER IN RELATION TO THE MI-GRATION OF THEIR COMPONENTS

BY

F. KOHLRAUSCH

Director of the Reichsanstalt, Charlottenburg

Presented May 6, 1876, before the Göttingen Academy of Sciences

(*Göttingen Nachrichten*, 1876, 213 ; Carl, *Repertorium*, **13**, 10, 1877)

CONTENTS

ON THE CONDUCTIVITY OF ELECTRO-LYTES DISSOLVED IN WATER IN RELATION TO THE MI-GRATION OF THEIR COMPONENTS

F. KOHLRAUSCH

I TAKE the liberty of presenting, as an appendix to a previous communication (these Proceedings, 1874, 405), a few remarks on the Mechanics of Electrolysis. I have shown with Mr. Grotrian, in the paper mentioned, that dilute aqueous solutions of the chlorides of all the alkalies and alkali earths possess nearly the same conductivity when an equal number of equivalents are dissolved. $_{\wedge} \epsilon_{j} \omega_{\lambda} \ \cdots \omega_{\lambda} \epsilon_{\lambda} \cdot \aleph_{\mathcal{U}} \overline{\iota_{\cdots}}$.

If the differences still remaining be compared with the transference numbers of the migrating components, as determined by Wiedemann, Weiske, and especially by Hittorf* in his classical work on *The Migration of Ions During Electrolysis*, an evident connection between the two quantities is at once noticed. By following this matter further, one is led to an assumption, remarkable for its simplicity, regarding the nature of the electrical resistance of dilute solutions, which I will now develop with the aid of the previous examples, as well as by some which I have more recently observed.

---

* Hittorf, Pogg. *Ann.*, **89**, 177; **98**, 1; **103**, 1; **106**, 513. Wiedemann, ibid., **99**, 182. Weiske, ibid., **103**, 466.

Pure water does not possess an appreciable conductivity, and hence it is most natural to regard current conduction in an aqueous solution of a body as due, not to conduction by the water, but by the dissolved particles. This view is probably held by most physicists at the present time.* According to it the water acts only as a medium in which the electrolytic displacements take place, and the electrical resistance of the solution would be the frictional resistance which the migrating elements of the salt, etc., experience against the water particles and against each other.

If, now, the solution be very dilute, this friction will occur for the most part on the water particles. Hence one will be further tempted to conclude—and this is a conclusion which to my knowledge has never previously been drawn — *that in a dilute solution every electrochemical element (e.g.,* hydrogen, chlorine, or also a radical, as NO_3) *has a perfectly definite resistance pertaining to it, independent of the compound from which it is electrolyzed.* As we know little concerning the nature of a solution, however, it is clear that such an assumption is justified only by experimental verification.

I think I can now prove that the facts correspond very nearly to the above law for a large group of substances—namely, for all the univalent acids and their salts whose conductivity has been investigated.

For this purpose let us consider dilute solutions which contain an equal number of electrolytic molecules in equal volumes. I shall call such solutions *electrochemically equivalent.* Of course the electrolytic molecule is not always to be regarded as the molecule now assumed in chemistry, but only that fraction of the latter which is decomposed by the same quantity of electricity as a molecule composed of two chemically univalent components.

Let each solution form a column of unit cross-section, and let it be acted on by an electric force (potential gradient), unity. If the ions have the opposite velocities, u_0 and u, then by Faraday's law, according to which each migrating partial molecule carries with it a quantity of electricity independent of its nature, the current is proportional to $u_0 + u$

* Compare, *e.g.*, Hittorf; Quincke, Pogg. *Ann.*, **144**, 2; Wiedemann, Galvanismus (2), I., p. 471.

(and to the number of molecules contained in unit length of the column, which, however, shall be the same in all solutions).

On the other hand, the strength of the current through unit cross-section due to an electromotive force unity is, as well known, nothing else than what is called the conductivity, l, of the solution, which must therefore be proportional to $u_0 + u$. The ratio of the velocities u_0 and u has been determined by Hittorf for a large number of solutions. We will call with Hittorf $n = \dfrac{u_0}{u_0 + u}$ the transference number of the component which has the velocity u_0.

Now let two electrochemically equivalent solutions of two compounds, I and II, be given, which have one component in common—e.g., that one having the velocity u_0, while the other components have the velocities u_1 and u_2 respectively. Let the corresponding conductivities of the solutions be l_1 and l_2. Then from the above

$$\frac{l_1}{l_2} = \frac{u_0 + u_1}{u_0 + u_2} = \frac{\left(\dfrac{u_0}{u_0 + u_2}\right)}{\left(\dfrac{u_0}{u_0 + u_1}\right)} = \frac{n_2}{n_1}.$$

Hence our hypothesis requires *that the conductivity of electrochemically equivalent solutions of two electrolytes, having a component in common, shall vary inversely as the transference numbers of the common component; or, that the product of the conductivity of each solution and the corresponding transference number of the common component shall be equal.*

This conclusion is verified in the following compilation of all material on electrolytes of univalent acids at my disposal.

TABLE

	l_1	n_1		l_2	n_2	$\dfrac{l_1}{l_2}$	$\dfrac{n_2}{n_1}$
KCl	977	0.510	NaCl	807	0.63	1.21	1.23
"	"	"	NH_4Cl	949	0.51	1.03	1.00
"	"	"	$Ca\frac{1}{2}Cl$	742	0.68	1.32	1.33
"	"	"	$Mg\frac{1}{2}Cl$	712	0.69	1.37	1.35
"	"	"	$Ba\frac{1}{2}Cl$	800	0.62	1.22	1.22
"	"	"	$Sr\frac{1}{2}Cl$	777	0.65	1.26	1.27
"	"	"	HCl	3230	0.161	0.302	0.316
KNO_3	927	0.495	$AgNO_3$	810	0.53	1.14	1.07
"	"	"	HNO_3	3360	0.142	0.275	0.287
KBr	1044	0.514	HBr	3100	0.178	0.329	0.346
KI	1048	0.50	HI	3190	0.258	0.328	0.516
KCl	977	0.490	KBr	1044	0.486	0.94	0.99
"	"	"	KI	1048	0.50	0.93	1.02
"	"	"	KNO_3	927	0.505	1.05	1.03
"	"	"	$KClO_3$	843	0.55	1.16	1.12
"	"	"	KAc	699	0.676	1.40	1.38

In the whole table there is but one considerable difference between the ratios of n and of l—namely, in the case of HI. But in this very case it is probable, from the values of the transference numbers which Hittorf gives, that u has been found too large for iodine. Such an error is easily possible, as only one observation was made, and as control experiments cannot be made at both electrodes in the case of acids as in the case of salts.

The assumption of the independent migration of ions may also be tested by the transference numbers alone, and hence confirmed or confuted in the case of substances whose conductivity is not yet known. It is easily seen that the following relation must hold between the transference numbers of the four compounds which can be formed from two pairs of electrochemical atoms A A′ and B B′:

Let $m_1\ n_1$. $m_2\ n_2$, $m_3\ n_3$, $m_4\ n_4$ be the transference numbers of the electrolytes A B, A B′, A′ B, A′ B′ respectively, where m refers always to A, n to B, and of course $m+n=1$.

Then our assumption evidently requires that

$$\frac{m_1}{n_1}\frac{m_4}{n_4} = \frac{m_2}{n_2}\frac{m_3}{n_3}.$$

In the following six examples, taken from Hittorf's determinations, together with Wiedemaun's values for HNO_3, the deviations from the required relation amount to scarcely more than is to be expected from the uncertainty of the observations themselves.

A	A'	B	B'	Cl n_1	H n_2	Na n_3	$K NO_3$ n_4	$\dfrac{m_1 m_4}{n_1 n_4}$	$\dfrac{m_2 m_3}{n_2 n_3}$
K	Na	Cl	NO_3	0.51	0.495	0.63	0.614	0.60	0.60
Na	Ba	Cl	NO_3	0.63	0.614	0.616	0.61	0.38	0.39
H	Ca	Cl	NO_3	0.161	0.142	0.68	0.62	3.18	2.84
K	Na	Cl	I	0.51	0.50	0.63	0.62	0.59	0.59
K	Na	Cl	Ac	0.51	0.324	0.63	0.443	1.21	1.23
K	Ag	Ac	NO_3	0.324	0.495	0.627	0.626	1.88	1.72

From the experimental confirmation of these two consequences, I am of the opinion that the law here suggested has great probability—that is to say, that we may speak of the mobility* of an electrolytic component in water.

I give below, as provisional, the following numbers for the ~~mobility~~ velocity u, referred to that of hydrogen as unity :

H	Br	Cl	I	K	NH_4	NO_3
1.00	0.19	0.19	0.18	0.18	0.18	0.17

Ag	ClO_3	Ba	Na	Ca	Sr	Mg	Ac
0.15	0.15	0.12	0.11	0.10	0.10	0.09	0.09

The mobility of hydrogen exceeds, therefore, that of the other elements five to eleven times, and it may indeed be asserted with certainty *that the high conductivity of acids is due to the fact that hydrogen is one of their migrating components.* Possibly the same remark also applies to the good conduction of the alkalies in solution.

The above numbers also give the possibility of calculating the conductivity of a dilute solution of an electrolyte the components of which have the mobilities u and u'. If one part by weight of the solution contains p parts by weight of the elec-

*["*Beweglichkeit.*" *In modern phraseology it is more customary to speak of the "velocity of migration," "Wanderungsgeschwindigkeit" of an ion.*]

trolyte, and if A denotes its electrochemical molecular weight, then the conductivity of the solution at 18° referred to mercury is given approximately by

$$k = 0.027 \,\frac{u+u'}{A} \cdot p.$$

The factor by which p is multiplied represents, therefore, the specific conductivity.

Finally the force which produces a definite velocity of one of the above components may be expressed in mechanical units, as first shown by W. Weber and R. Kohlrausch* in the case of water, although at that time under assumptions regarding electrolysis which do not correspond to ours. By introduction of the absolute resistance of mercury and of the electrochemical equivalent, one obtains for hydrogen, for example, the velocity $\dfrac{2.9 \text{ mm.}}{10^{12} \text{ sec.}}$, corresponding to an electrical force of separation unity expressed in absolute magnetic measure (millimeter, milligram, and second as fundamental units). From this it follows, that if the electromotive force of a Daniell cells acts on a column of dilute HCl (or HBr, HNO$_3$, etc.) a millimeters in length, the hydrogen will be moved along with a velocity of $0.33 \,\dfrac{\text{mm.}}{\text{sec.}}$. The velocity of any other ion under the same conditions is obtained by multiplying this number by u.

If the electromotive force be calculated in mechanical units on the assumption that the hydrogen is moved by the force exerted by the electromotive force on the quantity of electricity migrating with it, it is found that in order to force the hydrogen electrolytically through the water with a velocity of $1 \,\dfrac{\text{mm.}}{\text{sec.}}$, a force equal to the weight of 33,000 kg. must act on each milligram of hydrogen.† If 33,000 be divided by the product of the electrochemical molecular weight and the value of u for another component, the corresponding value for that component is obtained.

---

* *Abhandlungen der K. Sächs. Ges. d. Wiss.*, v., 270.

† These figures are based solely on resistances to conduction, and have, therefore, nothing to do with the overcoming of the forces of chemical affinity which manifest themselves in the polarization of the electrodes.

Further experiments alone can decide how far the laws here developed may be generalized on the one hand, or must remain limited to certain groups of substances on the other, and to what extent they apply rigidly or only approximately. I must mention here, however, that of the substances whose conductivities were investigated, one—namely, acetic acid—stands quite isolated from the above relations, if it be assumed, from analogy with the acetic acid salts, that hydrogen forms one of its ions. If this be so, acetic acid should be a very good conductor, while in reality, even in aqueous solution, it does not even approach the worst conducting of the solutions here considered. From this quite abnormal behavior it would follow that in acetic acid other conditions are present, either in regard to its chemical constitution, or in the nature of its solution in water,* than in the case of the other acids, or even the acetic acid salts. An exactly similar case occurs in aqueous ammonia, although this does not belong to the examples mentioned in this paper. I expected that aqueous ammonia would conduct particularly well, since ammonia salts conduct exceedingly well, and potassium and sodium hydrate, moreover, conduct much better than their salts. But, on the contrary, this substance, like acetic acid, is such a poor conductor that it evidently belongs to an entirely different class of bodies. This fact lends support to the view of some chemists, that aqueous ammonia does not contain the compound NH_4OH corresponding to the alkali hydrates, but is only a solution of NH_3. I defer the further discussion of such cases, as well as my observations on the polybasic acids and their salts, to the future, remarking only in the mean time that their conductivity comes out too large when calculated from the above transference numbers of their components.

In conclusion I wish to call attention to another noteworthy point of comparison between the conductivity and the transference numbers of dissolved electrolytes, to which Mr. Hittorf himself has very kindly called my attention. Most electrolytes investigated show a decreasing value of the transference number of the cation with increasing concentration. A few, how-

---

* Hydrocyanic acid appears to act similarly.

ever, retain nearly the same transference relations in strong as in dilute solutions. This is more or less the case with the potassium salts, and next to them ammonium chloride, the only ammonium salt investigated.

Now the conductivities of the last-mentioned substances show a similar agreement, in contrast to the others. In the case of most electrolytes the ratio of the conductivity to the percentage composition of the solutions diminishes continuously and very considerably; so much so, in fact, that not infrequently the known phenomenon of a maximum occurs. In the case of potassium and ammonium salts, however, this ratio is much more constant. From this relation, noticed as stated by Hittorf, it would follow that the resistances to motion which arise in denser solutions affect, in general, the cation more than the anion. But 1 would add at once, however, that a diminished mobility must also be ascribed to the latter, in order to explain the observed conductivity of stronger solutions.

WÜRZBURG, *May* 1, 1876.

BIOGRAPHICAL SKETCH

FRIEDRICH KOHLRAUSCH, son of Rudolph Kohlrausch the physicist, was born at Rinteln, Germany, in 1840. He was educated at the Polytechnicum at Cassel, and at the Universities of Marburg, Erlangen, and Göttingen, making his doctor's degree at the last-named university, under Wilhelm Weber in 1863. He then acted as assistant in the astronomical observatory at Göttingen, in the laboratory of the Physical Society at Frankfort, and in the University of Göttingen. He held the professorship of physics in the Polytechnicum at Zurich, 1870–71; at Darmstadt, 1871–75; and at the University of Würzburg until 1888, when he succeeded Kundt as director of the physical laboratory at Strassburg. On the death of v. Helmholtz, in 1894, he left Strassburg to accept the appointment of director of the Physikalisch Technische Reichsanstalt at Charlottenburg, which position he now holds.

The numerous contributions of Professor Kohlrausch to physical science have been mostly in the domain of electricity and magnetism, and are characterized by the high

degree of precision with which they have been car d out.
A number of the best methods and instruments n sed in
electrical and magnetic measurements are due to h:. In addition to the extensive series of researches on the electrical
conductivity of electrolytes, begun in 1868 and continued to
the present time, may be mentioned his investigations on elas-
ticity in 1866 ; his series of researches on magnetic measurements, 1881–84; and his determination with W. Kohlrausch,
of the electrochemical equivalent of silver in 1886. He is
also the author of one of the best-known works on physical
laboratory methods, entitled *Leitfaden der praktischen Physik*,
which has been translated into four different languages.

BIBLIOGRAPHY*

REFERENCE BOOKS

Kohlrausch and Holborn. *Leitvermögen der Electrolyte.* 1898.
This contains, besides theory and details of methods of conductivity measurement, compiled data on the conductivity of aqueous solutions recomputed in *reciprocal ohms*, tables of transference numbers and velocities of migrations of ions, and a very complete author index of work on electrical conductivity and related subjects.

Ostwald. *Lehrbuch der Allgemeinen Chemie,* vol. ii., part 1. 1892.
Faraday's Law, pp. 579–592; Transference Numbers, pp. 593–620; Kohlrausch's Law, p. 639.
Electrochemie. Ihre Geschichte und Lehre. 1896.

Wiedemann. *Lehre von der Electricität,* vol. ii. 1894.
Faraday's Law, pp. 467–475; Transference Numbers, pp. 572–586; Kohlrausch's Law and Absolute Velocity of Migration, pp. 926–933 ; Electrochemical Equivalent, vol. iv. 1898. pp. 726–736.

FARADAY'S LAW

Becquerel.   *Ann. de Chim. et de Phys.,* **66**, 91, 1837.
Buff.   *Ann. d. Chem. u. Pharm.,* **85**, 1, 1853 ; **94**, 15, 1855.
Despretz.   *C. R.,* **42**, 707, 1856.
Faraday.   *Exp. Researches, Ser.* V., VII., VIII.
Foucault.   *C. R.,* **37**, 580, 1853.
Gray.   *Phil. Mag.* (5), **22**, 389, 1886 ; (5), **25**, 179, 1888.
Jamin.   *C. R.,* **38**, 390, 1854.
de la Rive.   Pogg. *Ann.,* **99**, 626, 1856 ; *Ann. de Chim. et de Phys.* (3), **46**, 41, 1856.
Logeman and van Breda. *Phil. Mag.* (4), **8**, 465, 1854.
Matteucce.   *Ann. de Chim. et de Phys.,* **58**, 78, 1835 ; **74**, 109, 1840.
Ostwald and Nernst. *Zeit. f. Phys. Chem.,* **3**, 120, 1889.
Shaw.   *Rep. Brit. Assoc.,* 1886, p. 318.
                Phil. Mag. (6), **23**, 138, 1887.
Soret.   *Ann. de Chim. et de Phys.* (3), **42**, 257, 1854.

---

* No attempt has been made to make the bibliography given below complete. It is hoped, however, that none of the more important recent papers have been omitted.

LAWS OF ELECTROLYTIC CONDUCTION

ELECTROCHEMICAL EQUIVALENT

Heydweiller.	*Inaug. Dissertation*, Würzburg, 1886.
F. Kohlrausch.	*Gött. Nach.*, 1873.
	Pogg. Ann., **169**, 170, 1873.
F. and W. Kohlrausch.	*Sitz. ber. d. Phys. Med. Ges. z.*, Würzburg, 1884.
	Wied. Ann., **27**, 1, 1886.
Kahle.	*Zeit. f. Instrumentenkunde*, **17**, 144, 1897; **18**, 141, 1898.
Köpsel.	*Wied. Ann.*, **31**, 268, 1887.
Mascart.	*C. R.*, **93**, 50, 1881.
	Journ. de Phys. (2), **1**, 109, 1882.
	Ibid. **3**. 283, 1884.
Patterson and Guthe.	*Phys. Rev.*, **7**. 257, 1898.
Potier and Pellat.	*Journ. de Phys.* (2), **9**, 381, 1890.
	Lum. Elec., **32**, 84, 1889.
Rayleigh and Sidgwick.	*Phil. Trans.* (A), **2**, 411, 1884.
	Proc. Roy. Soc., **37**, 144, 1884.

TRANSFERENCE NUMBERS

Bein.	*Wied. Ann*, **46**, 29, 1892.
	Zeit. f. Ph. Ch., **27**, 1, 1898.
Campetti.	*At. Tor.*, **29**. 228, 1894; **32**. 1897.
	Nuov. Cim. (3). **35**, 225, 1894; (4), **1**, 73. 1895.
Cattaneo.	*Rend. Linc.* (5). **5**, II., 207, 1896; (5), **6**, I., 279, 1897.
Chassy.	*Ann. de Chim. et de Phys* (6), **21**, 241, 1890.
des Coudres.	*Wied. Ann.*. **57**. 232. 1896.
Gordon.	*Zeit. f. Ph. Ch.*. **23**. 469, 1897.
Hittorf.	*Pogg. Ann.*, **89**. 177, 1853; **98**, 1, 1856; **103**, 1, 1858; **106**. 338, 513, 1859.
	Ostwald's Klassiker der Exak. Wissens., Nos. 21, 23.
Hofgartner.	*Zeit. f. Ph. Ch.*. **25**, 115, 1898.
Kirmis.	*Wied. Ann.*. **4**, 503. 1878.
Kistiakowsky.	*Zeit. f. Ph. Ch.*. **6**, 105. 1890.
Kümmell.	*Wied. Ann.*, **64**, 655, 1898.
Kuschel.	*Wied. Ann.*, **13**, 289. 1881.
Lassana.	*Att. Ist. Ven.* (7). **3**. 1111, 1892; **4**. 1568, 1893.
	Riv. Scien. Indust. Firenze, **29**. 10, 1897.
Lenz.	*Mem. St. Pet. Ak.* (7), **30**, 64, 1882.
Löb and Nernst.	*Zeit. f. Ph. Ch.*, **2**, 948. 1888.
Mather.	*Johns Hopkins Univ.*, **16**. 45. 1897.
McIntosh.	*J. Phys. Chem.*, **2**, 273, 1898.
Rosenheim.	*Zeit. f. Anorg. Ch*. **11**, 220. 1893.
Schrader.	*Zeit. f. Elek. Ch.*. **3**, 498, 1896.
Weiske.	*Pogg. Ann.*. **103**. 466. 1858.
G. Wiedemann.	*Pogg. Ann.*, **99**. 177. 1856; **104**, 162. 1858.

LAWS OF ELECTROLYTIC CONDUCTION

VELOCITY OF MIGRATION

FOR tabulated values (referred to ohms as unit of resistance and 18° C.), see Kohlrausch & Holborn's *Leitvermögen der Electrolyte*, p. 201. Also Wiedemann's *Electricität*, vol. ii., pp. 927, 929.
For values at 25° C., see Ostwald's *Lehrbuch*, vol. ii., part 1, p. 675. See also F. Kohlrausch, *Gött. Nachr.*, 1876, p. 213; Wied. *Ann.*, **6**, 170, 1879; **26**, 213, 1885; **50**, 385, 1893; **66**, 785, 1898.

For values of the velocity of migration of organic anions and cations, see literature compiled by Kohlrausch & Holborn on the electrical conductivity of organic acids and bases by Ostwald, Bredig, and others.

ABSOLUTE VELOCITIES

Budde.	Pogg. *Ann.*, **156**, 618, 1875.
F. Kohlrausch.	*Gött. Nachr.*, 1876, 213.
	Wied. *Ann.*, **6**, 201, 1879 ; **50**, 402, 1893.
Lodge.	*Brit. Assoc. Rep.*, 1886, 389.
Nernst.	*Zeit. f. Elec. Chem.*, **3**, 308.
Sheldon and Downing.	*Phys. Rev.*, **1**, 51, 1893.
Weber.	*Zeit. f. Ph. Ch.*, **4**, 182, 1889.
Whetham.	*Phil. Trans.*. **184** A, 337, 1893 ; **186** A, 507, 1895.
	Zeit. f. Ph. Ch., **11**, 220, 1893.

INDEX

INDEX

THE END

TEXT-BOOKS IN PHYSICS

THEORY OF PHYSICS

By Joseph S. Ames, Ph.D., Associate Professor of Physics in Johns Hopkins University. Crown 8vo, Cloth, $1 60; by mail, $1 75.

In writing this book it has been the author's aim to give a concise statement of the experimental facts on which the science of physics is based, and to present with these statements the accepted theories which correlate or "explain" them. The book is designed for those students who have had no previous training in physics, or at least only an elementary course, and is adapted to junior classes in colleges or technical schools. The entire subject, as presented in the work, may be easily studied in a course lasting for the academic year of nine months.

Perhaps the best general introduction to physics ever printed in the English language. . . . A model of comprehensiveness, directness, arrangement, and clearness of expression. . . . The treatment of each subject is wonderfully up to date for a text-book, and does credit to the system which keeps Johns Hopkins abreast of the times. Merely as an example of lucid expression and of systematization the book is worthy of careful reading.—*N. Y. Press.*

Seems to me to be thoroughly scientific in its treatment and to give the student what is conspicuously absent in certain well-known text-books on the subject—an excellent perspective of the very extensive phenomena of physics. — Professor F. E. Beach, *Sheffield Scientific School of Yale University.*

A MANUAL OF EXPERIMENTS IN PHYSICS

Laboratory Instruction for College Classes. By Joseph S. Ames, Ph.D., Associate Professor of Physics in Johns Hopkins University, author of "Theory of Physics," and William J. A. Bliss, Associate in Physics in Johns Hopkins University. 8vo, Cloth, $1 80; by mail, $1 95.

I have examined the book, and am greatly pleased with it. It is clear and well arranged, and has the best and newest methods. I can cheerfully recommend it as a most excellent work of its kind.—H. W. Harding, *Professor Emeritus of Physics, Lehigh University.*

I think the work will materially aid laboratory instructors, lead to more scientific training of the students, and assist markedly in incentives to more advanced and original research.—Lucien I. Blake, *Professor of Physics, University of Kansas.*

It is written with that clearness and precision which are characteristic of its authors. I am confident that the book will be of great service to teachers and students in the physical laboratory.—Harry C. Jones, Ph.D., *Instructor in Physical Chemistry, Johns Hopkins University.*

NEW YORK AND LONDON
HARPER & BROTHERS, PUBLISHERS

STANDARDS IN NATURAL SCIENCE

COMPARATIVE ZOOLOGY

Structural and Systematic. For use in Schools and Colleges. By JAMES ORTON, Ph.D. New edition, revised by CHARLES WRIGHT DODGE, M.S., Professor of Biology in the University of Rochester. With 350 illustrations. Crown 8vo, Cloth, $1 80 ; by mail, $1 96.

The distinctive character of this work consists in the treatment of the whole Animal Kingdom as a unit ; in the comparative study of the development and variations of organs and their functions, from the simplest to the most complex state ; in withholding Systematic Zoology until the student has mastered those structural affinities upon which true classification is founded ; and in being fitted for High Schools and Mixed Schools by its language and illustrations, yet going far enough to constitute a complete grammar of the science for the undergraduate course of any college.

INTRODUCTION TO ELEMENTARY PRACTICAL BIOLOGY

A Laboratory Guide for High Schools and College Students. By CHARLES WRIGHT DODGE, M.S., Professor of Biology, University of Rochester. Crown 8vo, Cloth, $1 80 ; by mail, $1 95.

Professor Dodge's manual consists essentially of questions on the structure and the physiology of a series of common animals and plants typical of their kind—questions which can be answered only by actual examination of the specimen or by experiment. Directions are given for the collection of specimens, for their preservation, and for preparing them for examination ; also for performing simple physiological experiments. Particular species are not required, as the questions usually apply well to several related forms.

THE STUDENTS' LYELL

A Manual of Elementary Geology. Edited by JOHN W. JUDD, C.B., LL.D., F.R.S., Professor of Geology, and Dean of the Royal College of Science, London. With a Geological Map, and 736 Illustrations in the Text. New, revised edition. Crown 8vo, Cloth, $2 25 ; by mail, $2 39.

The progress of geological science during the last quarter of a century has rendered necessary very considerable additions and corrections, and the rewriting of large portions of the book, but I have everywhere striven to preserve the author's plan and to follow the methods which characterize the original work.—*Extract from the Preface of the Revised Edition.*

NEW YORK AND LONDON
HARPER & BROTHERS, PUBLISHERS